C0-AJY-740

Computer-Aided Design
of Magnetic Circuits

COMPUTER-AIDED DESIGN OF MAGNETIC CIRCUITS

ALEXANDER KUSKO AND THEODORE WROBLEWSKI

Research Monograph No. 55
The M.I.T. Press
Cambridge, Massachusetts, and London, England

Set in Press Roman by the Science Press Inc.
Printed and bound in the United States of America
by Colonial Press Inc.

Library of Congress catalog card number: 69–19249

To O. Howard Biggs

FOREWORD

This is the fifty-fifth volume in the M.I.T. Research Monograph Series published by the M.I.T. Press. The objective of this series is to contribute to the professional literature a number of significant pieces of research, larger in scope than journal articles but normally less ambitious than finished books. We believe that such studies deserve a wider circulation than can be accomplished by informal channels, and we hope that this form of publication will make them readily accessible to research organizations, libraries, and independent workers.

Howard W. Johnson

PREFACE

The material presented in the book was developed over a five-year period during which computer design programs were being prepared for the Magnetic Components Department of Sylvania Lighting Products, subsidiary of General Telephone and Electronics Corp. These computer programs were used to reduce design time for new transformers and lamp ballasts, to explore new magnetic circuit configurations, and to obtain more accurate designs to reduce cut-and-try work on prototypes. During the same period, the material was organized into notes and used as the basis for several series of lectures to the design group; each chapter of the book corresponds roughly with one of these lectures. In addition, specific topics were presented to the industry in technical papers.

We recognized early in our work that the mere transfer of hand design methods to computer programs would not be successful, particularly for magnetic circuits operating at high levels of saturation and for ballasts operating with highly nonlinear gaseous-discharge lamp loads. For such applications there were no accurate hand design methods; the industry traditionally relied upon a cut-and-try approach. Hence, much of our work was to develop analytical techniques and mathematical models for the magnetic circuits and the electric circuits in which they were placed, before even attempting to complete programs for computer-aided design.

The purpose of the book is to present our approach to modeling magnetic circuits, to handling the interface with the electric circuit; and to preparing the sequence of steps for computer-aided design.

Much of the material, particularly on regulating transformers and leakage-reactance structures is original work and should be of interest to all engineers concerned with magnetic-circuit analysis and design. The book is an account of the application of computers to the design of a class of equipment, starting with the problem; discussing the assumptions, equations, and models; and finally showing the flow diagrams and typical input and output forms in use.

The chapters of the book are arranged so that the simplest magnetic circuit device, the reactor, is considered first, followed by leakage-reactance transformers and regulating transformers. The last chapter covers conventional multiwinding transformers. For each class of device, the theory and modeling are treated, followed by numerical examples and our approach to computer-aided design. The examples are oriented toward lamp ballasts, because their magnetic circuit operation is complex and because most of the programs were prepared for their design.

The book will be useful for practicing engineers in the magnetics and lighting industries, engineers and programmers concerned with computer-aided design, and teachers and students alike who are interested in the application of computers to engineering problems. A list of references on computer-aided design of transformers and electric machines is given in the Appendix.

We should like to acknowledge the efforts of Mr. Joseph Fitzgerald and Mr. Edwin Zabierek of Sylvania, and of Mr. Bronic Szpakowski of Alexander Kusko, Inc. during the development of the material for this book.

ALEXANDER KUSKO
THEODORE WROBLEWSKI

CONTENTS

Computer-Aided Design
of Magnetic Circuits

1 INTRODUCTION

The digital computer provides a rapid and accurate means for analyzing and designing magnetic circuits, provided that the program accounts for the non-linear and other unusual behavior of the magnetic material and the circuit in which it is placed. The mere transfer of formerly used hand analytical and design techniques to the computer is inadequate to secure satisfactory results on magnetic circuit problems.

1.1 Magnetic Circuits

Magnetic circuits are combinations of ferromagnetic (iron) core members and current-carrying windings linking the cores. The cores are made from powdered iron, ferrite, solid iron, steel strips or punched laminations. The windings can be single or multiple coils, placed in various arrangements with respect to the other windings and cores.

Each magnetic circuit is designed and built to fulfill a function. The function can be to provide a given value of inductance, in which case the core usually has an air gap. The function can be to transform voltage, current, or impedance levels from one winding to another on a common core. The function can be to provide force or torque, in which case one core member or winding is allowed to move or rotate. Or, the function can be to provide special waveform shapes as in frequency multipliers and pulse generators; to control energy flow to a load as in magnetic amplifiers; or to regulate voltage or power by saturation of a section of the core.

The common denominator of magnetic circuit design is the management of the saturation characteristic of the core material and the dissipation of the heat generated in the windings and the core. The same laws of physics govern the circuits from the smallest to the largest.

1.2 Categories of Magnetic Design Problems

The computer is of particular value in four categories of magnetic design problems. They are:

1. That in which the magnetic circuit is straightforward but a computer program will save time and keep the parameters organized.
2. That in which the magnetic circuit is not straightforward, because of gaps or saturation, and a computer program makes an accurate analytical solution or design possible.
3. That in which the parameters of the magnetic circuit device are not known because the electric circuit in which the device operates is nonlinear. The computer is programmed to solve the electric circuit problem first, then proceed to the magnetic circuit.
4. That in which the design must be optimized for space, weight, or economic considerations and a computer is programmed to carry out many designs or to follow an optimization procedure.

1.3 Economics of Computer Usage

A strong economic basis underlies the use of computers for the design of magnetic circuits as well as other devices and systems.

A computer design program establishes a procedure which no longer relies upon a set of skills developed by the design engineer. The program is prepared and refined by the most skilled people available and their skills remain with the program after they are gone. The competent design engineer can apply his talents to preparing new programs or refining old ones while the computer carries out designs without him. The computer does not displace design engineers, but forces them to use their skills at a higher intellectual level.

A second feature of a computer is its use for exploring and optimizing designs. Optimization is essential economically; it achieves lowest cost, weight, and losses. It can be carried out by scanning over a range of input variables, or by using sophisticated mathematical techniques in multivariable space.

A third consideration is the use of the computer for the bread-and-butter aspect of the design, manufacturing, and selling of products. Since a computer can prepare a design quickly, it can provide information on material and labor costs for use in quotations, and can prepare complete design and manufacturing instructions for items for which orders have been received. A good computer program can produce accurate designs which eliminate cut-and-try prototype work and reduce the time to reach production.

A fourth and intangible feature of the use of a computer for design is the significant education provided to the engineer who formulates the mathematical models and programs for the computer. The result is both greater understanding of the operation of the device and improvement of its form and structure.

1.4 Background

Computer-aided magnetic-circuit design has been incorporated in programs for designing transformers and electric motors. Two classical papers describe the preparation and use of these programs, the paper on transformers by Williams, Abetti, and Mason* and on induction motors by Veinott.† Additional pertinent references are listed in the Appendix. The magnetic circuits in these examples are straightforward and are handled in a conventional manner.

The procedures described in the two references and in others as well were developed for use in the manufacturing industry. They reflect the problem of getting the most design work done and the most economical structures in the least time. The inputs to the programs are given in terms of the terminal performance requirements of the equipment to be designed. The designer retains considerable control over the program, in that he supplies some of the first estimates of the variables and selects one of several designs that the computer supplies. The designs are carried out in terms of standard dimensions, laminations, wire sizes and mechanical fittings.

Optimization of the design in the classical sense is not carried out because, first, the figure of merit (cost, weight, etc.) is a discontinuous function of the variables, and second, the practical requirements impose so many constraints that little freedom is left in the variables. Instead, one or two key variables are varied and the designer selects one of several computer outputs as the best design. Typical key variables are core area in a transformer, which governs the copper-to-iron ratio, and stack length in an induction motor.

1.5 Scope of Book

The purpose of the book is to expose the reader to the field of computer-aided design as it applies to magnetic circuits. The purpose of all computer-aided design is to facilitate building and selling of certain devices and equipment. Each area of design has specialized techniques and problems; we shall deal only with those associated with design of magnetic circuits.

The chapters of the book treat the categories listed in Section 1.2. Chapter 2 treats the air-gap reactor, both as the simplest unit of a magnetic circuit, and as a magnetic circuit complicated by an air gap. Computer design of re-

* S. B. Williams, P. A. Abetti, and H. J. Mason, "Complete design of power transformers with a large-size digital computer," *AIEE Trans.*, vol. 78, pt. III, pp. 1282-1291, February 1959.

† C. G. Veinott, "Syntheses of induction motor designs on a digital computer," *AIEE Trans.*, vol. 79, pt. III, pp. 12-18, April 1960.

actors follows in Chapter 3. Leakage-reactance transformers, more complicated gapped structures, are discussed in Chapter 4. Regulating transformers, which utilize saturation, are treated in Chapter 5. The operation of magnetic circuits in combination with other electrical devices is analyzed in Chapter 6; the example of the gaseous-discharge lamp ballast is used. The background of Chapters 4, 5 and 6 is utilized in Chapters 7 and 8, on the numerical and computer design of regulating transformers. Finally, the uncomplicated multiwinding transformer is used in Chapter 9 to show the use of a computer to handle a straightforward design procedure.

2 AIR-GAP REACTOR

The ac air-gap reactor or choke is the simplest useful form of magnetic circuit. It consists of at least one winding on a magnetic core; the core is usually gapped or made from powdered iron material to achieve a reasonably constant reactance. The ac reactor is used as an element of a tuned filter, as a means for loading equipment on test, for compensating transmission lines and high-voltage cables, and for ballasting gaseous-discharge lamps. The function of the reactor is to provide inductive reactance with an arbitrarily small variation of its parameters over the range of operating current.*

2.1 Reactor Geometry

Air-gap reactors are built in one of three geometries depending upon the application. Sketches of the structures are shown in Figure 2.1. The structure of Figure 2.1a is termed a shell-type reactor; the air gap is placed either in the center leg or in all three legs. The winding on the center leg links two parallel magnetic paths which tend to shield the winding from the external environment.

The second structure, shown in Figure 2.1b, is termed a core-type reactor. It can be built with one winding for the complete magnetic circuit or two equal windings connected in series or parallel. The two-winding structure reduces leakage from the core, keeps the average turn length down, and provides good exposure for heat dissipation from the windings.

The third structure, shown in Figure 2.1c, is termed a toroidal-core reactor. The core is made of powdered iron so that it acts like a distributed gap. The winding is continuously wound over the distributed gap. There is no external magnetic field produced.

* For more background material on air-gap reactors and magnetic circuits see M.I.T. Staff, *Magnetic Circuits and Transformers*, New York: John Wiley & Sons, 1943.

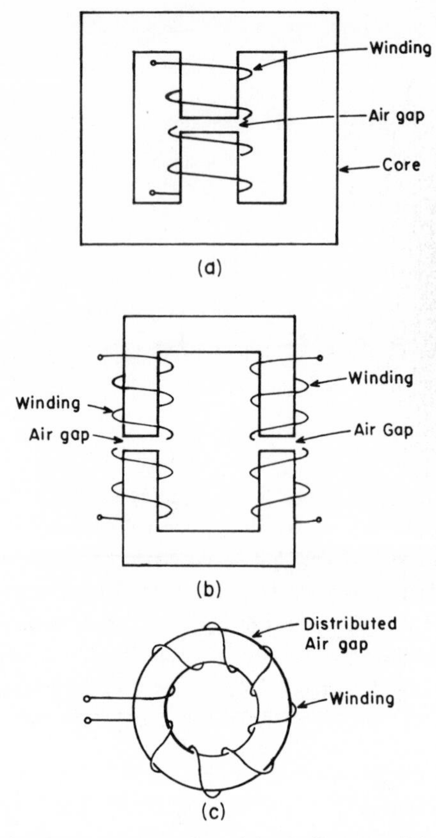

Figure 2.1 Typical reactor geometries: (a) shell type, (b) core type, (c) torodial.

Variations of the structures are used, but they all utilize the magnetic material to carry the magnetic potential of the winding to the faces of the air gap in which the energy is stored. Some structures are less expensive to manufacture, but they are all equivalent. An example of a reactor that was designed by computer is shown in Figure 2.2.

*2.2 Magnetic Circuit Model**

The model that will be used for the reactor is shown in Figure 2.3. It consists of a core of ferromagnetic material having a mean length l_m and effective cross-sectional area A_m. The material is assumed to have a single valued relationship between the flux density B_m and the corresponding field inten-

* All units are in the mks system unless specified otherwise. A table of symbols is given at the end of the book.

Figure 2.2 A lamp ballast reactor designed by computer.

sity H_m denoted by permeability $\mu_m = B_m / H_m = f(B_m)$. The core also has an air gap as shown, which may be located anywhere along the path. The air gap has a mechanical length g_o and an effective area of A_g. Finally, the core is wound with a coil of N turns by which the circuit current is coupled to the

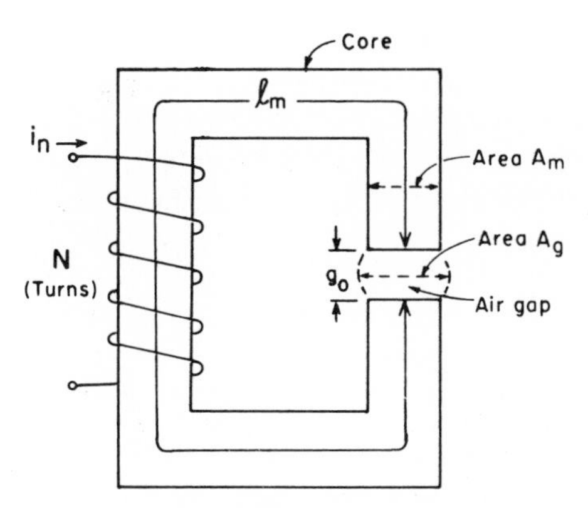

Figure 2.3 Model of reactor magnetic circuit.

magnetic circuit. The physical reactors can have a variety of geometries but they can all be described by the model.

The problem of analysis is to find the reactance and losses of the reactor for a given geometry, material, and operating current. On the other hand, the problem of design is to find the combination of core, air gap, winding, and wire size that meets the requirements of inductance, linearity, and current. The reactor must be the most economical to build within the limitations of size, temperature rise, and efficiency. Obviously, there is in infinite number of combinations to meet the requirements of inductance and current, but less as restrictions are added. The number of restrictions must be balanced against the number of design variables to insure that the reactor can be designed; the selection of variables is one of the first problems in preparing a computer program.

2.3 *Magnetic Circuit Equations*

The equations used for magnetic circuit analysis stem from Maxwell's equations applied to simple geometries in the quasi-static case. In integral form, they are simply

$$\oint \mathbf{H} \cdot d\mathbf{l} = \int_s \mathbf{J} \cdot \mathbf{n} \; da, \tag{2.1}$$

$$\int_s \mathbf{B} \cdot \mathbf{n} \; da = 0, \tag{2.2}$$

and

$$\oint \mathbf{E} \cdot d\mathbf{l} = -\frac{d}{dt} \int_s \mathbf{B} \cdot \mathbf{n} \; da. \tag{2.3}$$

If we assume for the present that the magnetic flux remains in the core and that the core cross-section is constant, then the flux density component along the length of the core is B_m. If we also assume that the flux density is constant at B_g along the air gap, then we can write Equation 2.1 as

$$\frac{B_m}{\mu_m} l_m + \frac{B_g}{\mu_0} g_0 = N i_n, \tag{2.4}$$

which is the familiar ampere-turn sum expression. Using Equation 2.2 for the interface between the core and the gap, we can write

$$B_m A_m - B_g A_g = 0. \tag{2.5}$$

The combined Equation 2.4 and 2.5 is

$$\frac{B_g}{\mu_0}\left(g_0 + \frac{A_g}{A_m}\frac{\mu_0}{\mu_m} l_m\right) = Ni_n. \tag{2.6}$$

Equation 2.6 shows that the magnetic circuit model can be treated as a core of infinite permeability with an equivalent gap of length $g' = g_0 + \Delta_g$, where the gap increment is given by

$$\Delta_g = \frac{A_g}{A_m}\frac{\mu_0}{\mu_m} l_m. \tag{2.7}$$

The ratio μ_m/μ_0 is the relative permeability k_m of the iron at flux density B_m. Hence, as the flux density is allowed to increase, the value of k_m decreases, and the equivalent gap increases, thereby tending to reduce the inductance.

2.4 *Air-Gap Correction*

The magnetic field crossing the air gap tends to bulge or fringe as shown in Figure 2.3 so that the flux density B_g in the gap is less than the value B_m in the core. A suitable correction is to assume that the cross-sectional area of the gap is obtained by increasing each dimension of the gap by the physical gap length g_0. If the core has outer dimensions of width x and depth d, then the cross-sectional area of the gap is

$$A_g = (x + g_0)(d + g_0). \tag{2.8}$$

The gap length g_0 is usually kept small compared to x and d to reduce the extent of the fringing. The fringing flux component that enters and leaves the core through the flat faces of the lamination produces eddy currents in the laminations and additional heating. Hence gaps are kept small or broken up into a succession of gaps in series to reduce the fringing in actual physical reactors.

2.5 *Inductance*

The quasi-static behavior of magnetic circuits and the calculation of inductance is easily carried out using the parameter of permeance $\mathcal{P}$. The permeance is analogous to the conductance of a resistive electric circuit, and is the reciprocal of the magnetic circuit reluctance $\mathcal{R}$. The permeance for a magnetic circuit of total flux ϕ and ampere-turns NI is defined as

$$\mathcal{P} = \frac{\phi}{NI}. \tag{2.9}$$

For a magnetic circuit of constant cross-section A, the flux density B along the path ℓ is constant and the permeance becomes

$$\mathcal{P} = \frac{\phi}{NI} = \frac{BA}{Hl} = \mu \frac{A}{l}. \tag{2.10}$$

A coil of N turns linking a magnetic circuit of permeance $\mathcal{P}$ has an inductance L of

$$L = \frac{N\phi}{I} = N^2 \mathcal{P}. \tag{2.11}$$

Hence, the permeance is the one-turn inductance of a magnetic circuit. The inductance of the model of Figure 2.3 in terms of the equivalent gap g' is

$$L = N^2 \mu_0 \frac{A_g}{g'}. \tag{2.12}$$

For several magnetic circuits in parallel linking the same coil, the total permeance is the sum of the circuit permeances. For example, a shell-type core having gaps g_1' and g_2' in the outer legs will have an inductance of

$$L = N^2 \mathcal{P}_t = N^2 \left(\mu_0 \frac{A_{g1}}{g_1'} + \mu_0 \frac{A_{g2}}{g_2'} \right). \tag{2.13}$$

To obtain inductance in henrys for dimensions in meters, use $\mu_0 = 4\pi \times 10^{-7}$ of the mks system. For dimensions in inches use

$$\mathcal{P} = \frac{A_g \,(\text{in}^2)}{g' \,(\text{in})} (3.195 \times 10^{-8}). \tag{2.14}$$

For example, the inductance of a 300-turn coil on a core having gap dimensions $A_g = 2.0$ in^2 and $g' = 0.050$ in is

$$L = (300)^2 \left(\frac{2}{0.050} \right) (3.195 \times 10^{-8}) = 0.115 \text{ H}.$$

2.6 Gap Energy

A more general method for establishing the relationship in a reactor between the physical and electrical parameters is through the magnetic-field energy stored in the air-gap volume. The energy stored in a unit volume of free space, such as in an air gap, is given by

$$w = \frac{1}{2} \frac{B_g^2}{\mu_0}. \tag{2.15}$$

In mks units, the energy density w is given in joules (watt-seconds) per cubic meter for $\mu_0 = 4\pi \times 10^{-7}$ and B_g expressed as webers per square meter (1 weber = 10^8 lines). If the permeability of the medium is other than μ_0, the same expression applies provided that the permeability is constant during the time that the magnetic field is raised from zero to the final value. Otherwise, the energy density must be found from the integral form

$$w = \int_0^{B_g} H(B)dB. \tag{2.16}$$

The total energy W stored in a gap of uniform field B_g and dimensions A_g and g' is thus

$$W = \frac{1}{2}\frac{B_g^2}{\mu_0}(A_g g'). \tag{2.17}$$

The peak energy* W_p stored in the gap can be equated to the expression for energy in terms of the reactor inductance L and the peak current I_{np} that produces B_g

$$W_p = \frac{1}{2}LI_{np}^2 = \frac{1}{2}\frac{B_g^2}{\mu_0}(A_g g'). \tag{2.18}$$

For a sinusoidal current of frequency f, $I_{np} = \sqrt{2}\,I_n$ and $X_L = 2\pi fL = \omega L$, we find that

$$I_n^2 X_L = \omega W_p = \frac{1}{2}\omega\frac{B_g^2}{\mu_0}(A_g g'), \tag{2.19}$$

or,

$$P_q = \omega\left(\frac{1}{2}\frac{B_g^2}{\mu_0}\right)\mathcal{V}_g = \omega w \mathcal{V}_g. \tag{2.20}$$

Equations 2.19 and 2.20 are basic to all ac reactor designs. The reactive power rating P_q is equal to ω times the peak stored energy. The maximum design value of B_g is determined by the saturation flux density of the adjacent core material. Hence, at a given angular frequency ω, the reactive power rating $P_q = I_n^2 X_L$ of a reactor is directly proportional to the volume $\mathcal{V}_g$ of the air-gap space. The manner in which the volume is realized from an area A_g and length g' merely determines the relative amounts of iron and copper in the reactor, but at constant reactive power P_q. For $\omega = 377$ rad/sec and B_g in kilolines per square inch, the reactive power density is given by

$$\frac{P_q}{\mathcal{V}_g} = 0.59\, B_g^2 \text{ vars/in}^3. \tag{2.21}$$

*Since this will be a function of I_p, it should be correctly termed the co-energy. See H. H. Woodson and J. R. Melcher, *Electromechanical Dynamics*, New York: John Wiley & Sons, 1968, Chap. III.

For example, a reactor is required to have a reactance of 56.8 Ω at 3.2 A. Assume a flux density of B_g = 100 kilolines per square inch. The required volume of the air gap is

$$\mathcal{V}_g = \frac{580}{0.59 \times 10^4} = 0.098 \text{ in}^3.$$

For an air gap of $g' = 0.050$ in, the gap area is $A_g = 1.96 \text{ in}^2$.

2.7 Leakage Fields

The assumption was previously made that all of the magnetic flux remains in the core, except when the flux crosses the air gap. Actually, the flux does leak from the core, because the core permeability is not infinite, and finds paths through the air and other ferromagnetic objects, such as brackets, bolts, and cans. The leakage flux has two effects on a reactor; it produces increased losses and noise when it passes through the external ferromagnetic paths, and it adds to the calculated inductance by adding permeance to the air-gap path. From a design standpoint, we wish to select the design configuration that minimizes the stray field and to compensate the gap design for the leakage permeance so that the net inductance of the structure designed by the computer will be correct. The stray field is driven through the external paths by the magnetic potential of the core. The magnetic potential can be calculated along the surface of the core with respect to some reference point, just as the electric potential can be calculated along the surface of the wire of an electric circuit. Conceptually, the magnetic potential at any point along the core is equal to the total ampere-turns produced by the winding current to the point, minus the potential drop produced by the flux and the reluctance of the core.

A useful way of looking at the magnetic potential of a magnetic circuit, such as a reactor, is to plot the magnetic potential profile. Two plots are shown in Figure 2.4 for a gapped core with an N-turn winding. In the left-hand figure, the coil is mounted away from the gap. The magnetic potential plot is started at the center of the coil, where the potential is assumed to be zero, and the potential is traced clockwise around the core.

The potential rises by one half the coil ampere-turns to point a, less by the potential drop of the flux in the iron to point a. The potential continues to drop to point b in the iron, then sharply to point c in the air gap and finally to point d. The lower half of the coil then raises the potential back to zero at point o.

The potential profile for the right-hand core in Figure 2.4 is quite different. The potential drops from point o to a as the flux passes through the iron, and from d to o. However, the potential rises sharply by one-half the coil ampere-turns from a to b, drops through the gap fron. b to c, and rises again by the coil from c to d. The major rises and drops are confined to the region from a to d.

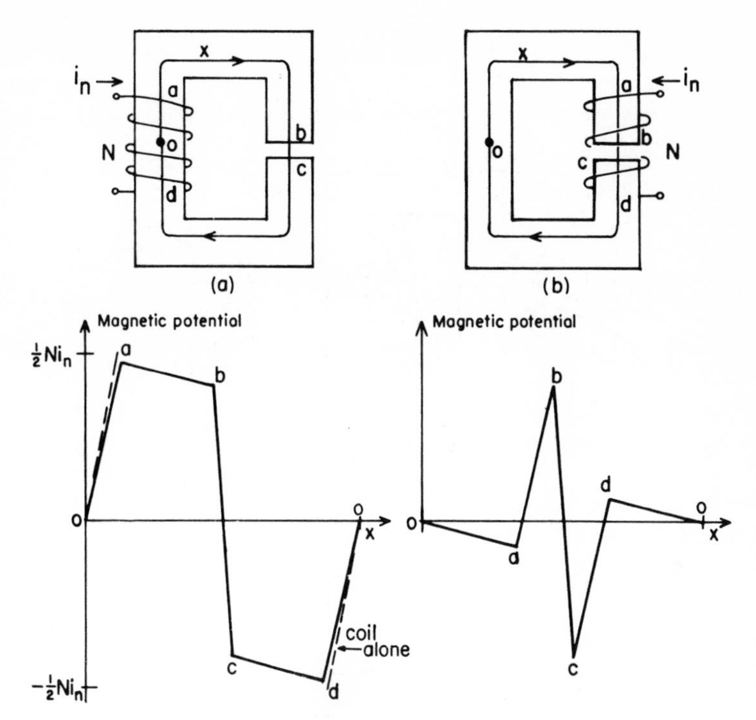

Figure 2.4 Magnetic potential profiles for reactors with (a) winding opposite gap, and (b) winding over gap.

The leakage fields produced by a magnetic circuit are proportional to the magnetic potentials between various parts of the core and the areas exposed. In the left-hand core, the horizontal limbs are at magnetic potential difference of almost Ni_n and will produce considerable stray field. The limbs of the right-hand core are at small potential difference and will produce minor stray fields. The stray field patterns are shown in Figure 2.5. The potential plots show the desirability of always placing the gaps under the coil that produces the ampere-turns for the gap.

The component contributed by the leakage flux will be termed the leakage inductance. The permeance contributed by the leakage flux can be calculated from the geometry of the reactor, and included in a computer design program.

The leakage flux paths for a core-type reactor or one half of a shell-type reactor are shown in Figure 2.6. There should be very little flux around the outside of the coil because the gap in the center of the coil consumes all of the coil ampere-turns. The field distribution should be somewhat as though the return limbs of the core appeared as two disks around the upper end and lower faces of the coil, keeping the external field small.

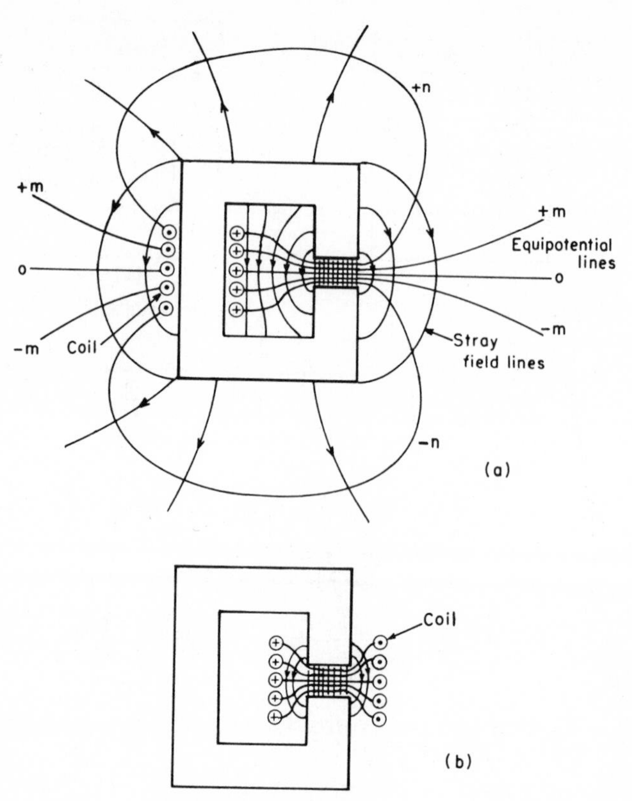

Figure 2.5 Equipotential and stray field lines for reactors of Figure 2.4.

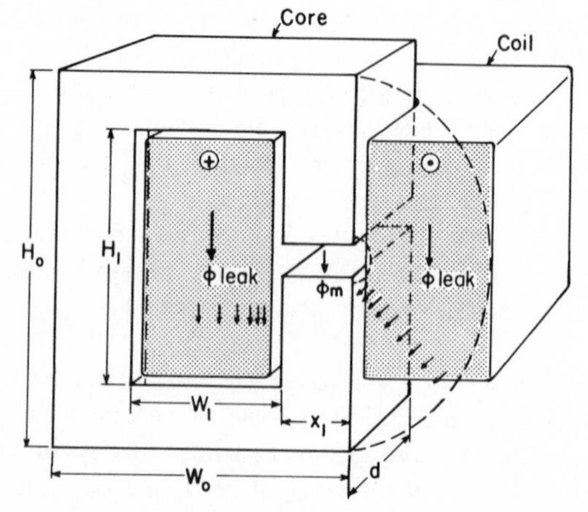

Figure 2.6 Main and leakage flux in core-type reactor.

The permeance for the leakage flux in the window is of the form

$$\mathcal{P}_{lw} = \frac{1}{3}\frac{W_1 d}{H_1}\mu_0, \tag{2.22}$$

where the 1/3 factor accounts for the fact that the leakage flux is threading the coil turns that are also providing the potential for the leakage flux. The geometrical parameters are shown in the figure.

The total permeance for the leakage flux will be taken as though the path were like that in the window but extended all around the coil. The total leakage permeance is thus

$$\mathcal{P}_l = \frac{1}{3}\left[\frac{(2W_1 + x_1)(2W_1 + d) - x_1 d}{H_1}\right]\mu_0$$

$$= \frac{2W_1\,\mu_0}{3H_1}(2W_1 + x_1 + d). \tag{2.23}$$

For all dimensions in inches and the desired inductance in henrys, $L = N^2\,\mathcal{P}_l$, the equation becomes

$$\mathcal{P}_l = \frac{2}{3}\frac{W_1}{H_1}(2W_1 + x_1 + d)(3.195 \times 10^{-8})$$

$$= 2.13 \times 10^{-8}\,\frac{W_1}{H_1}(2W_1 + x_1 + d). \tag{2.24}$$

A refinement such as this is ordinarily not required in a hand design procedure, but must be included in a computer program to insure that the inductance of the structure will meet the input specification.

2.8 Equivalent Circuits

Electrical equivalent circuits for magnetic circuits are extremely useful in studying the effect of individual parts of the magnetic circuit on the whole and of determining the operation of the magnetic circuit when connected to an external electric circuit.

The air-gap reactor has a useful equivalent circuit which can be determined from Equations 2.4 and 2.11. The inductance can be expressed as

$$L = \frac{N\phi}{i_n} = \frac{N^2 B_m A_m}{\dfrac{B_m}{\mu_m} l_m + \dfrac{B_g}{\mu_0} g_0} \tag{2.25}$$

$$= \frac{1}{\dfrac{l_m}{N^2 A_m \mu_m} + \dfrac{g_0}{N^2 A_g \mu_0}}.$$

Each of the denominator terms is in the form of $1/L$, so that we can say

$$L = \frac{1}{\dfrac{1}{L_m} + \dfrac{1}{L_g}}, \tag{2.26}$$

where L_m is the inductance of the core and L_g is the inductance of the gap. The equivalent circuit is shown in Figure 2.7. The two inductance elements

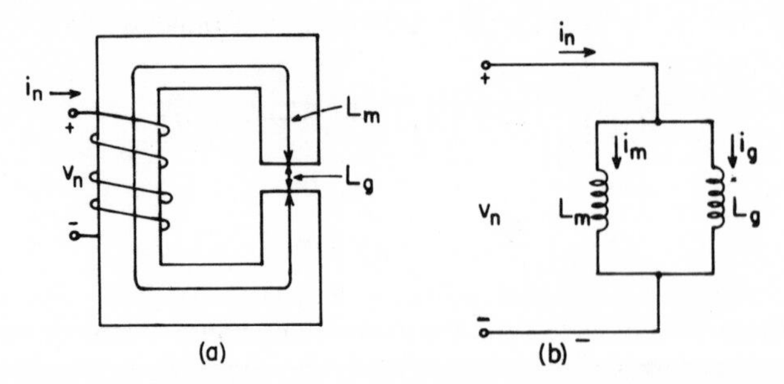

Figure 2.7 Equivalent circuit of reactor (a) shown at (b).

are in parallel although the magnetic circuit elements are in series. The equivalent circuit in terms of inductance is actually the circuit dual of the magnetic circuit. The currents in the two inductance elements represent the ampere-turns for each portion of the magnetic circuit and the voltage represents the flux in the core.

2.9 *Summary*

The air-gap reactor is a basic unit of magnetic circuits. The permeance of the magnetic circuit is the one-turn inductance, while the reactive power rating is proportional to the peak energy stored in the air gap. More complicated magnetic circuits can be treated as combinations of simpler air-gap and closed-core structures.

3 COMPUTER DESIGN OF REACTOR

We will consider the simplest category of design, in which the electrical parameters of a reactor are given and the independent design variables are reduced to a minimum by specifying as many as possible. The equations have been developed in Chapter 2. However, we wish to obtain a highly accurate design, so that when the reactor is built it will meet all of the specifications and require no cut-and-try modification.

3.1 Design Objective

The objective is to design an ac reactor consisting of a gapped core and a winding to meet the following electrical specifications:

Reactance at frequency $f = X$ ohms;

Current at X ohms $= I_n$ amperes.

In addition, the temperature rise on the surface must not exceed a value ΔT degrees at the operating condition. The structure will be the core type of Figure 2.6 or the shell type of Figure 3.1. We can consider reactance and inductance interchangeably because the frequency is given. Obviously, there are innumerable designs which will meet the electrical and thermal specifications. Let us examine the problem more closely. Four approximate relationships provide some insight into the variables that must be assigned numerical values to give the design. Using Equations 2.6 and 2.11, neglecting the effect of the core, we obtain

$$L \approx N^2 \mu_0 \frac{A_g}{g_0}, \tag{3.1}$$

$$N I_n \approx \frac{B_g}{\mu_0} g_0 . \tag{3.2}$$

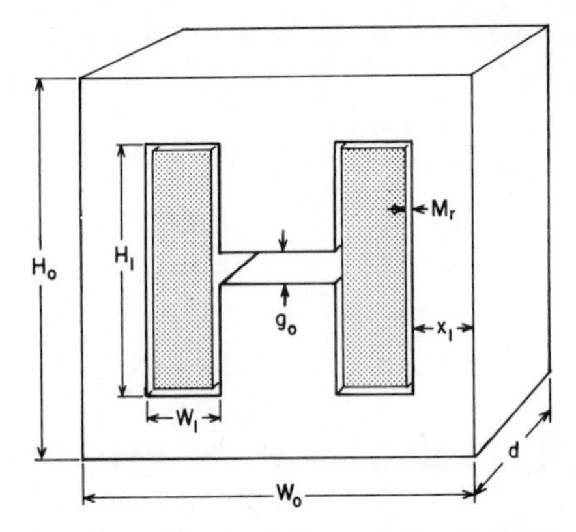

Figure 3.1 Configuration and dimensions of shell-type reactor.

In addition, the area of the window A_w is related to the current density J, neglecting the space factor, as

$$NI_n \approx JA_w, \tag{3.3}$$

and the temperature rise is related to all of the dimensions and densities as,

$$\Delta T = f\ (J, B_g, A_w, A_g). \tag{3.4}$$

We have four relationships and six variables that have not been specified, namely, A_g, g_0, N, B_g, J, and A_w. The variables L, I_n, and ΔT have been specified. We can either specify two more variables or write two more relationships. For example, we can specify minimum cost or weight as an additional relationship, or linearity of inductance with current.

From a practical standpoint, we will obtain a more useful design that is easier to manufacture if we specify several of the other dimensions and the physical gap length g_0. We can also let the temperature rise ΔT be a dependent variable, which is calculated from the finished design and used for iteration on an independent variable. We find that it is convenient to specify B_g, J, g_0, and some of the dimensions, and let the computer find N, A_g, A_w, and ΔT. Then the value of J can be raised or lowered to achieve the desired ΔT. The value of g_0 controls the copper-to-iron ratio and is a useful independent variable to scan for minimum cost or weight.

3.2 Inputs-Outputs

The inputs to a program to design the reactor are divided into four parts:

Tables

1. Wire table giving size, weight, resistance and cost by sizes
2. Core material B-H curves and losses
3. Temperature rise as a function of loss density

Electrical requirements
1. Reactance, X
2. Frequency, f
3. Current, I_n

Mechanical requirements
1. Core height, H_0
2. Core width, W_0
3. Strip width, D
4. Temperature rise, ΔT
5. Coil margin, M_r

Designer's parameters
1. Flux density, B_w
2. Current density, J
3. Physical gap, g_0
4. Core type, shell or core

The program must find the structure that satisfies the requirements and read out the dependent variables and other pertinent data, for example,
1. Turns
2. Winding Construction
3. Losses
4. Cost
5. Temperature rise
6. Core dimensions
7. Gap correction for saturation
8. Leakage factor
9. Wire size
10. Material weight

For the specified mechanical requirements, the core dimensions are the core build and the window width and height.

A typical input sheet is shown in Table 3.1 and an output sheet in Table 3.2 for a 400-W lamp ballast reactor design.

3.3 *Procedure*

The flow diagram for the procedure is shown in Figure 3.2. The procedure represents one of the simplest magnetic circuit routines but has two corrective loops to insure high accuracy.

The first loop is called the alpha loop, where A or α is a leakage factor. Of the total magnetic flux in the outer core limbs, the fraction A is assumed to go through the air gap and $(1-A)$ through the air paths and through the winding itself. A value of A is assumed, then calculated from the final geometry of the core, and fed back into the loop.

Table 3.1 Typical Printout of the Input Quantities of a Reactor Design for a 400-W Lamp Ballast

MERCURY LAMP CHOKE - SHELL TYPE - E ZABIEREK

IBEG	ILIM
10	21

NO	SQI	CUW	CUR	DIA	CC
10	0.0081	0.0319	0.0010	0.1071	0.7250
11	0.0064	0.0253	0.0012	0.0957	0.7265
12	0.0051	0.0200	0.0015	0.0855	0.7315
13	0.0040	0.0159	0.0020	0.0765	0.7335
14	0.0032	0.0126	0.0025	0.0684	0.7345
15	0.0025	0.0100	0.0031	0.0613	0.7450
16	0.0020	0.0079	0.0040	0.0548	0.7620
17	0.0016	0.0063	0.0050	0.0492	0.7825
18	0.0012	0.0050	0.0063	0.0440	0.8055
19	0.0010	0.0040	0.0080	0.0395	0.8190
20	0.0008	0.0031	0.0101	0.0353	0.8415
21	0.0006	0.0025	0.0127	0.0317	0.8625
22	0.0005	0.0019	0.0162	0.0284	0.8800
23	0.004	0.0015	0.0203	0.0255	0.9040
24	0.0003	0.0012	0.0256	0.0229	0.9500
25	0.0002	0.0009	0.0323	0.0206	0.9760
26	0.0001	0.0007	0.0410	0.0185	1.0115
27	0.0001	0.0006	0.0514	0.0165	1.0620
28	0.0001	0.0005	0.0653	0.0148	1.1160
29	0.0000	0.0003	0.0812	0.0134	1.1735
30	0.0000	0.0003	0.1037	0.0120	1.2175

A	-0.30441131E 01	B	0.44363111E-01	3.6070
A	-0.45718917E 01	B	0.59379629E-01	5.7917
A	-0.30758805E 01	B	0.45399621E-01	
A	-0.75823736E 00	B	0.10239686E-01	
A	0.10071278E 03	B	0.82821154E 00	

CJ INT	CJ END	DELTA CJ
2000.000	2000.000	1.000

BS INT	BS END	DELTA BS	EN	F	P
115.000	130.000	5.000	135.00	60.000	400.000

G INT	G END	DELTA G	VIN	VMR	IRON COST
0.060	0.075	0.015	3.200	0.300	0.182

HO	D
3.625	1.750

EO INT	EO END	DEO			
240.000	277.000	37.000			
0.925	120.535	60.267	130.178	158.366	169.446
49.489	131.275	4.849			

542.2299 VOLT-AMPERES

Table 3.2 Typical Printout of the Design of a Reactor for a 400-W Lamp Ballast. Design is Case No. 1 for the Inputs of Table 3.1

```
400W 240V CHOKE F19-91900  MAY 20, 1967
CASE NO.   1
CJ =  2000.0    BS =  115.00    BW =   81.19    GO = 0.0600   EO = 240.00

BUILD =  0.411

WIRE NO              17
TOTAL TURNS         286
TURNS PER LAYER      37
NUMBER OF LAYERS      7.75
COPPER-RESISTANCE =  0.99 OHMS          WEIGHT = 1.25 LBS     COST = $ 0.97
                                        WEIGHT = 5.73 LBS     COST = $ 1.04
                                                        TOTAL COST = $ 2.02

CORE AND SHUNT DIMENSIONS
    HO         WO         XL         HL         WL         D         DP
  3.625      4.176      0.750      2.125      0.588      1.750     2.684

TRUE SHORT CIRCUIT CURRENT = 7.239 AMP              POWER FACTOR = 0.551

COPPER LOSS = 10.20 WATTS            IRON LOSS = 13.58 WATTS

   GAP    WATTS PER LB  SURFACE AREA   TOTAL WATTS LOSS  TEMPERATURE RISE
 0.0600      1.183          60.70            23.79              46.36

         XT            VL1        EFF. GAP           EL
        49.48         131.2       0.062430          169.4
```

The second loop is called the AN loop and accommodates the correction of the air-gap areas A_g for the fringing field. The area A_g is set equal to the core area A_m for the first round, then corrected by the core dimensions and fed back into the loop as an iterative procedure. Both loops are allowed ten tries to reach stability; otherwise the program continues.

The correction Δg of the gap length for saturation of the core does not require a loop because the flux density is an input variable and the magnetic path length is known from the mechanical specifications.

A third loop can be added to compare the calculated value of ΔT with the specified value and to correct the input value of current density J. The procedure shown leaves the correction of J to the designer.

3.4 Design Steps

By way of example, a series of design steps is given which can be converted to suitable computer language and format to design a reactor. The steps follow the flow diagram of Figure 3.2.

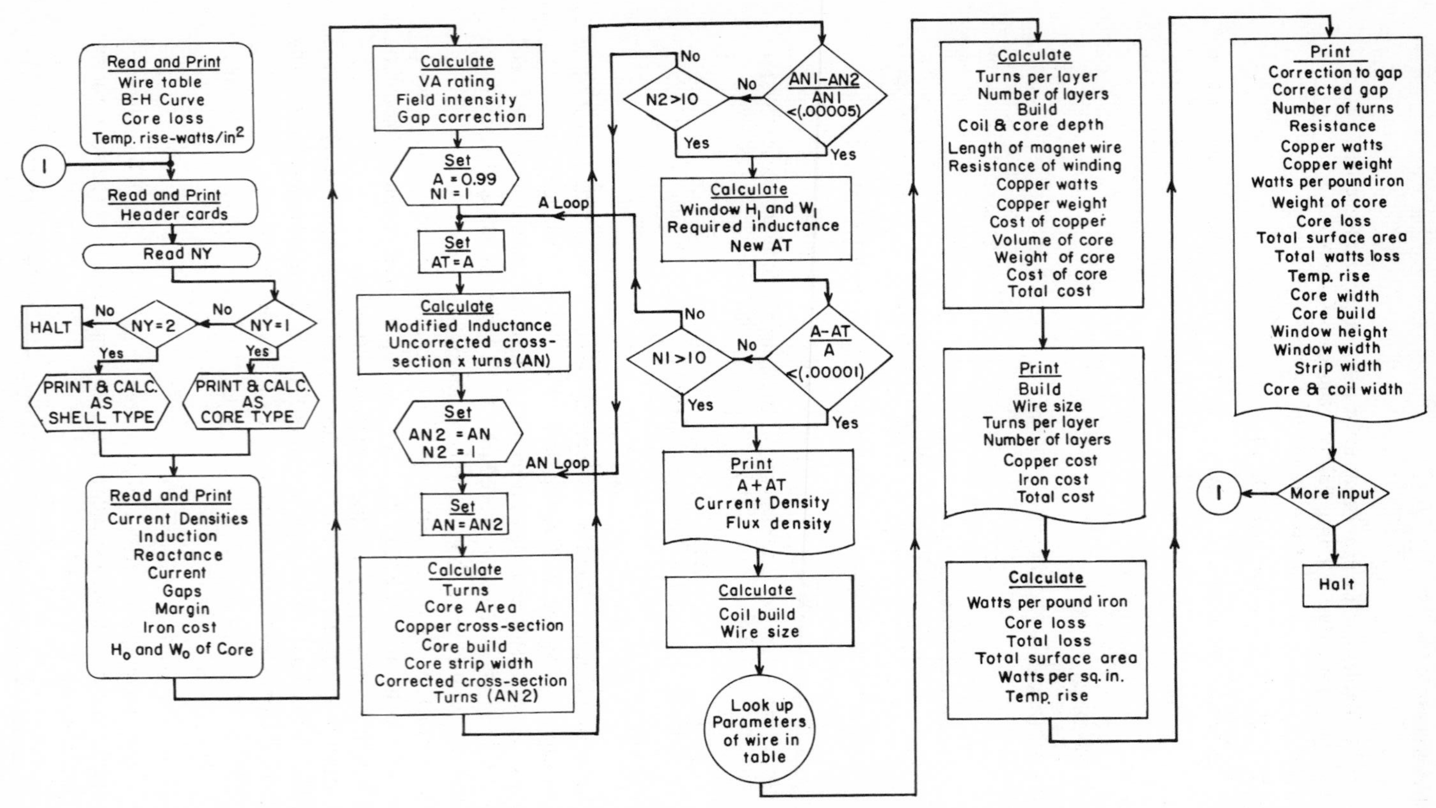

Figure 3.2 Flow diagram of computer program for reactor design.

The units are inches, millihenrys, ampere-turns per inch, kilolines, and kilolines per square inch.

1. Volt amperes

$$VR = EL \times VIN, \tag{3.5}$$

where

$$EL = \text{reactor voltage},$$

$$VIN = \text{reactor current}.$$

2. Permeability

$$US = (BW/HS) \times 313.5, \tag{3.6}$$

where

$$BW = \text{working flux density},$$

$$HS = \text{field intensity from steel curve}.$$

3. Gap correction

$$G1 = 1.5 \times (H0 + W0)/US \quad \text{(core type)}. \tag{3.7}$$

$$= 1.5 \times (H0 + .5\ W0)/US \quad \text{(shell type)}.$$

where

$$H0 = \text{core height},$$

$$W0 = \text{core width}.$$

4. Total equivalent gap

$$GP = G0 + G1, \tag{3.8}$$

where

$$G0 = \text{mechanical gap}.$$

5. Modified inductance

$$GL = A \times VL, \tag{3.9}$$

where

$$A = \text{leakage factor, assumed};$$

$$VL = \text{specified inductance}.$$

6. Area-turn product

$$AN = EL \times A \times 10^5/(BW \times .94 \times 4.44 \times F), \tag{3.10}$$

where

$$F = \text{frequency},$$

$$.94 = \text{core material space factor}.$$

$$AN = AN1\ (AN\ \text{loop}).$$

7. Number of turns

$$PN = GL \times GP \times 10^5/(3.195 \times AN). \tag{3.11}$$

8. Gross core area

$$AS = AN/PN. \tag{3.12}$$

9. Gross copper area

$$ACU = PN \times (VIN/CJ) \times 1.15, \tag{3.13}$$

where

$$CJ = \text{current density},$$
$$1.15 = \text{factor for precision winding.}$$

10. Gross core build

$$CB = (W0 + H0)/4 \qquad \text{(core type)},$$
$$CB = (.5\ W0 + H0)/4 \qquad \text{(shell type)}.$$

$$\begin{aligned} XL1 &= CB - \sqrt{(CB)^2 - (W0 \times H0 - ACU)/4} \qquad \text{(core type)},\\ XL1 &= CB - \sqrt{(CB)^2 - (.5\ W0 \times H0 - ACU)/4} \qquad \text{(shell type)}. \end{aligned} \tag{3.14}$$

11. Core build

$$XL = XL1 - VMR, \tag{3.15}$$

where

$$VMR = \text{coil margin in window.}$$

12. Core strip width

$$\begin{aligned} D &= AS/XL \qquad \text{(core type)},\\ D &= .5\ AS/XL \qquad \text{(shell type)}. \end{aligned} \tag{3.16}$$

13. Corrected area-turn product

$$\begin{aligned} AN2 &= AN\ [\ (D + G0) \times (XL + G0)\ ]\ /AS \qquad \text{(core type)},\\ AN2 &= AN\ [\ (D + G0) \times (2XL + G0)\ ]\ /AS \qquad \text{(shell type)}. \end{aligned} \tag{3.17}$$

If $|AN1 - AN2|/AN1 < 5 \times 10^{-5}$, go to next step.

If $> 5 \times 10^{-5}$, return to AN (step 6), where AN = AN2.

14. Window height

$$H1 = H0 - 2XL. \tag{3.18}$$

15. Window width

$$\begin{aligned} W_1 &= W0 - 2XL \qquad \text{(core type)},\\ W_1 &= .5\ W0 - 2XL \qquad \text{(shell type)}. \end{aligned} \tag{3.19}$$

16. Winding width

$$WW1 = W1 - VMR. \tag{3.20}$$

17. Required inductance
The inductance required from the air-gap portion of the reactor

$$SL = VL - 2.13 \times W1 \; [(2 \times W1 + XL + D)/H1 \times 10^5] \; PN^2 \quad \text{(core type)},$$

$$(3.21)$$

$$SL = VL - 2.13 \times W1 \; [(2 \times W1 + 2XL + D)/H1 \times 10^5] \; PN^2 \quad \text{(shell type)}.$$

18. Check leakage factor

$$AT = SL/VL \tag{3.22}$$

If $|A - AT|/A < 10^{-5}$, go to next step.

If $> 10^{-5}$, return to A (step 5), where A = AT.

19. Reactor coil build

$$W2 = XL + 1.57\,W1 \quad \text{(core type)}, \tag{3.23}$$

$$W2 = 2\,XL + 1.57\,W1 \quad \text{(shell type)}.$$

20. Wire size

$$PWS = VIN/CJ. \tag{3.24}$$

Select wire size from table, at least area PWS.

21. Turns per layer

$$NPLP = WW1/DIA(I). \tag{3.25}$$

Round off to integral number of turns TPLP.

22. Number of layers

$$PLN = PN/TPLP. \tag{3.26}$$

23. Build of coil

$$Z1 = PLN \times DIA(I) \times 1.045. \tag{3.27}$$

24. Core and coil width

$$DP = D + 2\,Z1 + .187. \tag{3.28}$$

25. Length of magnet wire (feet)

$$PWL = (DP - .0935 + D + W2 + XL)\,PN/12 \quad \text{(core type)},$$

$$PWL = (DP - .0935 + D + W2 + 2\,XL)\,PN/12 \quad \text{(shell type)}. \tag{3.29}$$

26. Resistance of winding

$$PR = PWL \times CUR(I), \tag{3.30}$$

where

$CUR(I)$ = wire resistance per foot from table.

27. Copper loss

$$PCW = PR \times VIN^2. \tag{3.31}$$

28. Copper weight

$$PW = PWL \times CUW(I), \tag{3.32}$$

where

$CUW(I)$ = wire weight per foot from table.

29. Copper cost

$$COSTC = PW \times CC(I) \tag{3.33}$$

where

$CC(I)$ = wire cost per pound from table.

30. Volume of core

$$\begin{aligned} VMC &= 2 \times (H0 + W1) \times XL \times D \quad \text{(core type)},\\ VMC &= 4 \times (H0 + W1) \times XL \times D \quad \text{(shell type)}. \end{aligned} \tag{3.34}$$

31. Weight of core

$$WMC = VMC \times .25944, \tag{3.35}$$

where

$.25944$ = pounds per cubic inch of steel.

32. Cost of core

$$COSTI = WMC \times FECOS, \tag{3.36}$$

where

$FECOS$ = iron cost per pound.

33. Total cost of copper and iron

$$TCOST = COSTC + COSTI. \tag{3.37}$$

34. Core loss

$$CLMC = WMC \times WPP \times 2, \tag{3.38}$$

where

WPP = watts per pound at flux density BW,

2 = factor for gap loss.

35. Total losses

$$TEST = PCW + CLMC. \tag{3.39}$$

36. Power loss density

$$WD = TEST/S9, \tag{3.40}$$

where

$$S9 = \text{total surface area.}$$

37. Temperature rise

DT obtained from curve for convection cooling as a function of WD.

3.5 Summary

One particular procedure is shown in this chapter to achieve the design of an air-gap reactor. Other procedures can be prepared to meet different input requirements. For example, the design could be required to meet a maximum value of losses, to use a given size and shape of lamination, or to meet a maximum change of inductance with current. The procedure for the design of more complicated structures that are required to provide reactance can be reduced to the procedure for one or more simpler air-gap reactors, once the parameters for the reactors have been established. High accuracy is assured in the procedure in this chapter by correcting for the core saturation, the gap fringing, the air-path leakage flux and the gap losses.

4 LEAKAGE-REACTANCE TRANSFORMERS

Leakage-reactance transformers provide both transformation and relatively high reactance between the source and load. They are used for current limiting such as in welding, and arc-type loads, for decoupling the output circuit as in constant-voltage transformers, and for starting and ballasting gaseous-discharge lamps. The leakage-reactance transformer combines in one magnetic circuit the transformer and the air-gap reactor.*

4.1 Structure

Leakage-reactance transformers are defined as those having a means of providing an artificial leakage flux path between the primary and secondary windings. Leakage flux is driven through the path by the magnetic potential of the windings and produces a voltage in the windings equivalent to an inductive-reactance voltage drop in series with the load. The ratio of that voltage drop to the load current is the leakage reactance of the transformer, referred to the load circuit.

Various types of leakage-reactance structures are shown in Figure 4.1 as examples of methods for providing the leakage flux path. In some cases, the leakage path appears to link one of the windings and in other cases both of the windings. For a main core of very high permeability, it would make no difference where the leakage path is placed; for finite permeability, there are small differences. Another point of difference is the sharing of the main core by the leakage and mutual fluxes for some structures, and the relegation of the fluxes to separate paths, in other cases. For a regulating transformer, the

*T. Wroblewski and A. Kusko, "Computer design analysis of the leakage-reactance ballast," *Illum. Eng.*, vol. 61, no. 3, pp. 285–289, April 1966.

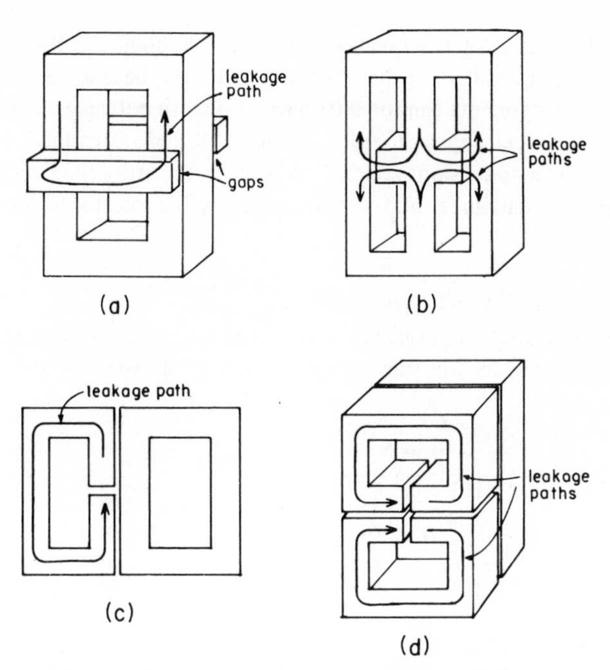

Figure 4.1 Leakage-reactance transformer core construction: (a) bar shunt; (b) X-core; (c) single leakage core, (d) double leakage core.

main core is driven down in permeability by saturation and there is a difference where the leakage path is placed.

The bar-shunt structure of Figure 4.1a is used for main cores of strip material, such as C-cores and formed cores, because the leakage flux has to leave and enter the main core through the narrow edge of the strip. Otherwise, the leakage flux will produce eddy currents and high losses. The structure is inexpensive to build and the gaps are easy to control, but the location of the gaps away from the coils produces relatively high stray fields. The structure is used for core and shell-type transformers.

The X-core structure of Figure 4.1b is commonly used in core or shell-type cores made from laminations, since the edges of the laminations face each other across the windows. The leakage-path shunt is either incorporated in the lamination, or is formed by pushing a small packet of laminations into the intercoil space. Practically all fluorescent ballasts and constant-voltage transformers use a structure of this kind. The gap dimensions are more difficult to control than in the other structures, the stray field is appreciable in certain directions and the geometry is fixed by the lamination sizes available.

The leakage-core structures of Figures 4.1c and d are used for cores of strip material. The net core cross-section under the winding with the leakage core may be greater than that in the bar-shunt structure because the mutual and leakage flux components cannot shift back and forth between the cores as the load shifts between open-circuit, short-circuit and rated-load conditions. However, the gap dimensions are easy to control, the stray field is much lower than all other structures, and the geometry is flexible for meeting special space requirements.

4.2 Magnetic Circuit Operation

The magnetic circuit conventions to be used for more than one winding and core are illustrated by the single magnetic circuit shown in Figure 4.2. The

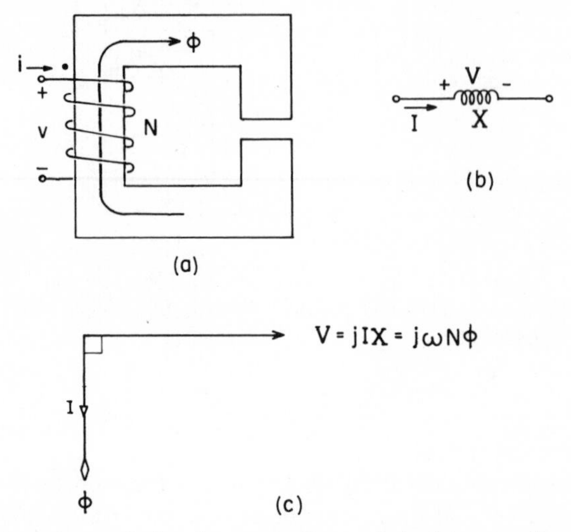

Figure 4.2 Polarity and circuit conventions of single magnetic gapped core: (a) reactor model; (b) electric circuit; (c) phasor diagram.

dot convention on the winding indicates that a current i into a dotted terminal produces a flux ϕ in the core in the direction shown, out of that end of the coil. A positive $d\phi/dt$ produces a voltage of positive polarity on the dotted terminal of the winding of value $Nd\phi/dt$, and of all dotted terminals on all of the windings on the core.

The equivalent circuit at the fundamental frequency of a lossless core and coil is merely the reactance X shown in Figure 4.2b. The reactance $X = 2\pi fL$ and the current I and voltage V are the rms values. The phasor diagram of the circuit is shown in Figure 4.2c. The voltage $V = jIX = j\omega N\phi$ leads the current I by 90°; the flux ϕ is in phase with the current I in accordance with the dot convention of Figure 4.2a.

The X-core structure is one in which the leakage flux and mutual flux share the same magnetic core under the windings. The magnetic circuit of such a structure is shown in Figure 4.3a. For simplicity, the transformer is shown supplying a resistive load R. The flux components are shown as ϕ_1 linking the primary N_1-turn winding, ϕ_2 linking the N_2-turn winding, and ϕ_l as the leakage flux in the two gaps.

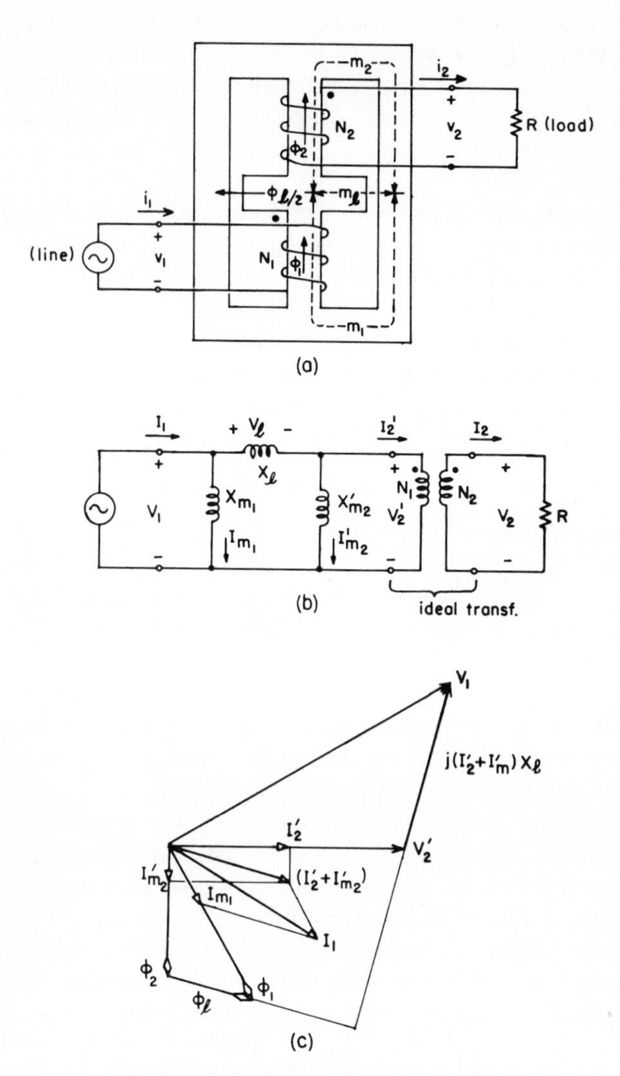

Figure 4.3 Leakage-reactance transformer: (a) magnetic-circuit model; (b) equivalent circuit; (c) phasor diagram.

The equivalent circuit for the structure is shown in Figure 4.3b.* The circuit is divided into a pi-reactance network and an ideal N_1-to-N_2 turn-ratio transformer. The reactance X_{m1} is the magnetizing reactance for the core around the path m_1; the reactance X_{m2} is the magnetizing reactance for the path m_2 referred to the N_1-turn winding. The reactance X_l is the leakage reactance corresponding to the permeance of the two paths m_l including the gap referred to the N_1-turn winding. The voltages across each of the reactances are equal to ωN_1 times the flux components in the paths corresponding to the reactances, that is, $V_1 = \omega N_1 \phi_1$, $V_2' = \omega N_1 \phi_2$, $V_l = \omega N_1 \phi_l$. The currents through the reactances correspond to the ampere-turns required for each of the fluxes, that is, $I_{m1}N_1$ for ϕ_1, $I_{m2}'N_1$ for ϕ_2, and $(I_2' + I_{m2}')N_1$ for ϕ_l. For a shell-type core, the flux paths m_l in the two halves of the core should be considered in parallel.

The phasor diagram for the structure is shown in Figure 4.3c. The magnetizing current components I_{m1} and I_{m2}' appear in quadrature with the voltages and in phase with the flux components ϕ_1 and ϕ_2. The current components are actually part of the line current I_1 and provide the magnetic potential for the flux components. The reactances X_{m1} and X_{m2} are calculated for the flux-density levels in the core or are measured on a structure as a function of the terminal voltages. The reactances can be referred to either side of the ideal transformer and placed wherever convenient in the equivalent circuit.

4.3 Autotransformer Operation

The various leakage-reactance structures shown in Figure 4.1 can be used with an autotransformer connection between the secondary and primary winding, as well as the isolated connection shown in Figure 4.3. The secondary winding can be connected to either a tap on the primary winding, the end of the primary winding, or an extension of the primary winding. The selection of the connection method depends upon the relationship of the line voltage to the required open-circuit and operating load voltage. A commercial line of leakage-reactance autotransformers for a specific lamp ballast may employ an extended primary winding for 120-V line operation, a direct-connected primary winding for 240-V line operation and a tapped primary winding for 480-V line operation. In all cases the secondary winding and core construction would remain relatively unchanged.

A diagram showing a leakage-reactance autotransformer with tapped primary winding is given in Figure 4.4a. The open-circuit voltage presented to the load is the sum of the tap voltage and the voltage induced magnetically in the secondary winding. Under load, the open-circuit voltage is reduced by the drop in the series reactance of the autotransformer.

*H. W. Lord, "An equivalent circuit for transformers in which nonlinear effects are present," *AIEE Trans.*, vol. 78, pt. I, pp. 580–586, 1959.

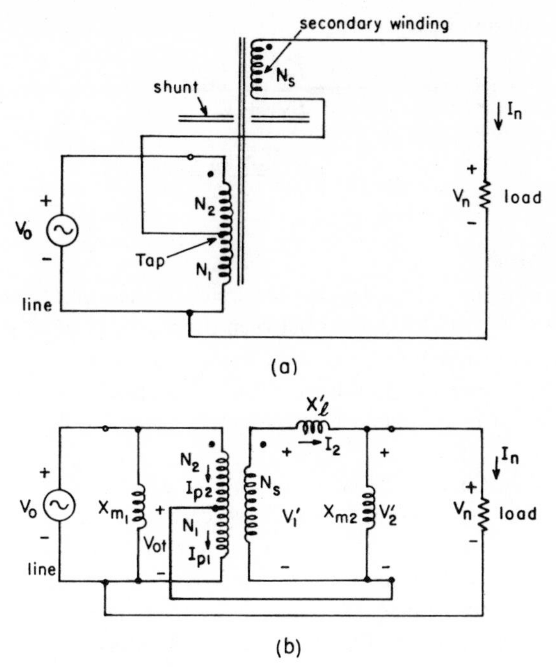

Figure 4.4 Leakage-reactance autotransformer: (a) circuit connection; (b) equivalent circuit.

The equivalent circuit for the leakage-reactance autotransformer of Figure 4.4a is shown in Figure 4.4b. The equivalent circuit is obtained from that of Figure 4.3b by reflecting the magnetizing reactance X_{m2} and the series reactance X_l through the turns ratio of the ideal transformer to the secondary winding. The ideal transformer is then tapped at the same percentage point on the primary winding as the actual transformer to permit the connection as shown in the equivalent circuit. The equivalent circuit can then be used for analysis and design of the autotransformer. The voltage-dividing action of the reactances X_l' and X_{m2}' on the secondary voltage corresponds to the diversion of a portion of the primary flux through the leakage shunts. The extended-primary-winding leakage-reactance autotransformer can be represented in the same fashion by an appropriate winding on the ideal transformer of the equivalent circuit. The magnetizing reactance X_{m1} does not affect the operation of the load circuit but merely draws reactive current from the line.

The winding currents for the tapped autotransformer of Figure 4.4b are determined from the ampere-turn and node equations

$$I_{p2}N_2 + I_{p1}N_1 = I_2N_s, \tag{4.1}$$

$$I_{p2} = I_{p1} + I_n. \tag{4.2}$$

From which we obtain

$$I_{p2} = \frac{I_2 N_s + I_n N_1}{(N_1 + N_2)}, \tag{4.3}$$

$$I_{p1} = \frac{I_2 N_s - I_n N_2}{(N_1 + N_2)}. \tag{4.4}$$

4.4 Reactance

The determination of the leakage reactance of the transformer requires the identification of all leakage paths between the windings. The series reactance shown as X_l in Figure 4.3b is the component of principal interest since it performs the decoupling function for the load. The magnetizing reactances X_{m1} and X_{m2} are not of particular interest except when the structure is used as a regulating transformer.

We showed in Chapter 2 that inductance or reactance could be related to stored magnetic field energy. The series reactance of the equivalent circuit of Figure 4.3b is related to or describes the energy stored in the magnetic fields in the air spaces of the structure. The magnetizing reactances are related to the energy stored in the magnetic fields in the core material. If the core material has an infinite permeability, then the energy stored is zero and the magnetizing reactances are also infinite.

The series reactance is composed of two principal components, namely, the component contributed by the energy stored in the shunt-path air gaps, and the component contributed by the energy stored in the air-path leakage fields through the core windows and around the outside of the core. If the design of a leakage-reactance transformer is based only on the air-gap reactance, then the total series reactance of the physical transformer will come out too high. For a computer design procedure, the reactance components must be very carefully calculated and accounted for to insure that the final design will have the required amount of reactance with the minimum size structure.

The inductance component contributed by the air gaps is given by the expression

$$L_g = N_1^2 \mathcal{P}_g, \tag{4.5}$$

where $\mathcal{P}_g$ is the permeance of the leakage shunt path including the air gaps, and N_1 is the primary winding turns. The permeance for the structure of Figure 4.3a is given by

$$\mathcal{P}_g = \mu_0 \frac{A_g}{g'}, \tag{4.6}$$

where $\mu_0 = 4\pi \times 10^{-7}$ for mks units and 3.195×10^{-8} for dimensions in inches. The term A_g is the effective area of the two gaps and should be corrected for fringing by adding the mechanical gap length g_0 to each dimension. That is, if the dimensions of each shunt are x wide and d deep, the effective area for two gaps is

$$A_g = 2(d + g_0)(x + g_0). \tag{4.7}$$

The gap length g' is the value of the mechanical gap length g_0 corrected for saturation of the shunt path. The correction is given as

$$g' = g_0 + \frac{l_m}{k_\mu} \frac{A_g}{A_m}. \tag{4.8}$$

The term $k_\mu = \mu_m/\mu_0$, the relative permeability of the shunt at flux density B_m. The term l_m is the length of the shunt path from center to center of the vertical legs. The correction term is quite small at moderate flux densities. However, when the secondary terminals of the transformer are short-circuited, such as when it is acting to limit load current, all of the primary core flux ϕ_1 is forced into the shunts and the flux density reaches its highest value. The effect of the high flux density is to reduce k_μ and to increase the equivalent gap g'. The reactance decreases allowing the load current to rise beyond its design value. The current under the short-circuit condition must be checked during the design process, or used as the design starting point, where it is important. The term k_μ is calculated from the B-H curve for the particular core material being employed.

The air-path leakage reactance of the transformer can be calculated from the dimensions of the structure and the numbers of turns by using the flux paths and components shown in Figure 4.5. These components can be defined as follows:

ϕ_{w1} = total leakage flux across the windows linking the primary and secondary windings. This includes the fringing flux shown in the top view of the autotransformer.

ϕ_{w2} = total leakage flux across the windows in the space between the windings, including the fringing flux.

ϕ_e = external leakage flux, which links the windings outside the windows as shown in the side view of Figure 4.5.

Each of these leakage flux components has associated with it a magnetic permeance and an inductance component referred to one of the two windings. In accordance with the location of the series reactance in Figure 4.3b, the inductance components will be referred to the primary winding. A bar-shunt transformer will have all three leakage flux components. The X-core will not have the ϕ_{w2} component because the shunt occupies the space between the windings.

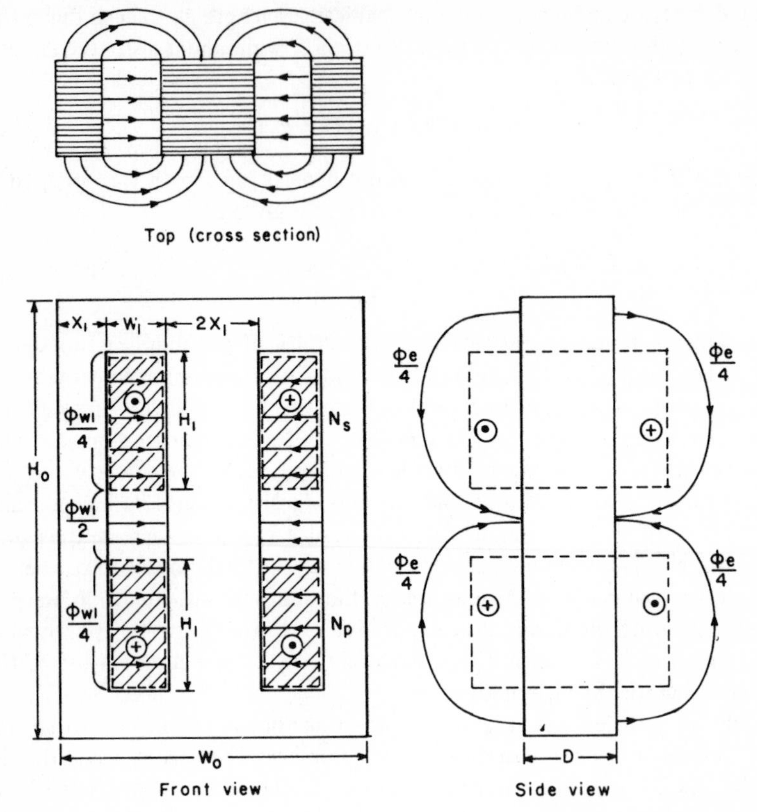

Figure 4.5 Air-path leakage flux components in leakage-reactance transformer.

The leakage flux component ϕ_{w1} across the windows and linking the windings has one portion that is uniform in the d-W_1 plane over part of the window and a second portion represented by the fringing field. The component of leakage inductance corresponding to these portions, referred to the primary winding, is

$$L_{w1} = \frac{4}{3} N_1^2 \mu_0 \frac{(d - X_1/2)H_1}{W_1} + \frac{4}{3} N_1^2 \mu_0 \frac{\pi H_1}{\cosh^{-1} \dfrac{(W_1 + X_1)}{X_1}}, \tag{4.9}$$

where the parameters are shown in Figure 4.5.

The leakage-inductance component for the flux ϕ_{w2} in window between the windings, referred to the primary winding, is

$$L_{w2} = 2N_1^2\mu_0 \frac{d(H_0 - 2X_1 - 2H_1)}{W_1}. \tag{4.10}$$

The leakage-inductance component for the flux ϕ_{w3} outside the window and linking the windings individually can be approximated by the expression

$$L_e = 8N_1^2\mu_0 (W_1 + X_1). \tag{4.11}$$

The total leakage inductance is then the sum of the components,

$$L_s = L_{w1} + L_{w2} + L_e. \tag{4.12}$$

The total series reactance for the leakage-reactance structure is thus given by the sum of the air-gap reactance and the air-path reactance. The expression is

$$X_l = 2\pi f(L_g + L_s). \tag{4.13}$$

The reactance can be considered constant. However, the air-gap inductance L_g is a function of the flux density in the shunt and will decrease for the short-circuit condition. The maximum impact on the total reactance is seldom more than a 20 per cent decrease for the short-circuit condition. The air-path inductance L_s is constant for all operating conditions.

4.5 *Phasor Diagrams*

The phasor diagram shows the manner in which the mutual and leakage fluxes combine in the core limbs during normal and short-circuit operation. The phasor diagram for the tapped leakage-reactance autotransformer of Figure 4.4 is shown in Figure 4.6. The secondary winding voltage under open-circuit conditions is V_2; under load, the voltage declines by the reactance drop jI_nX_l'' to the value V_2'. The load voltage V_n is thus the tap voltage V_{0t} plus V_2'.

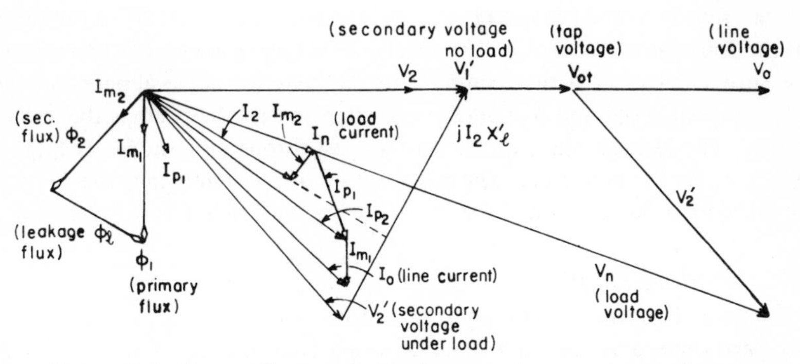

Figure 4.6 Phasor diagram for leakage-reactance autotransformer.

Each voltage is produced by a flux component: secondary flux ϕ_2 for V_2'; leakage flux ϕ_l for jI_2X_l'; and primary flux ϕ_1 for the line voltage V_0. Each flux component has a corresponding current which produces it: I_{m1} and I_{m2} for the core fluxes and I_2 for the leakage flux. The secondary core flux is seen to be less than the primary core flux. In a transformer having the same primary and secondary core areas, the secondary flux density would be lower than the primary flux density. In the regulating transformers to be described in Chapter 5, the opposite condition is produced by the capacitor.

The phasor diagram for the transformer of Figure 4.4 is shown with the output terminals short-circuited in Figure 4.7. The condition corresponds to

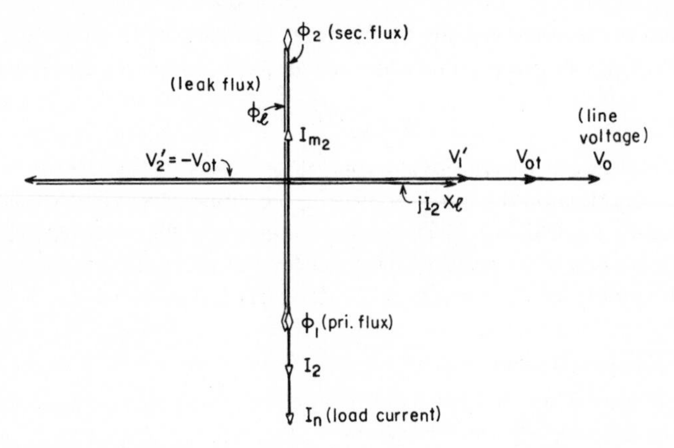

Figure 4.7 Phasor diagram for leakage-reactance autotransformer with output terminals short-circuited.

severe current limiting. For a transformer used as a lamp ballast, it corresponds to the lamp warmup time. The sum of V_2' and V_{0t} is zero. Actually, the magnitude of the tap voltage V_{0t} and the induced voltage V_1' is impressed on the leakage reactance X_l'. The reactance voltage can rise to several times the normal value. The flux components illustrate the physical phenomenon. The components ϕ_1 and ϕ_2 oppose each other so that their sum is the leakage flux ϕ_l. The leakage flux, which must all pass through the shunts, is approximately twice the core flux. The cross section of the shunts must be designed for this condition rather than the normal operating condition of Figure 4.6.

4.6 *Winding Tapping*

Tapping the primary winding of an autotransformer yields certain beneficial performance results and manufacturing conveniences, but can also increase the rating and size, depending upon the tap point. The circuit diagram for a tapped autotransformer is shown in Figure 4.4a. The tap point can also be located on an extension of the primary winding.

The primary winding is tapped for one of at least four reasons. First, the primary winding can be wound in two or more sections for multiple line-voltage operation. A typical arrangement is a winding for 120 V, 208 V and 240 V, where the tap point is at 120 V. Second, the tapping of the primary winding adds one additional degree of freedom to the design of the secondary winding and the leakage path gaps. For example, a transformer designed for 240-V line voltage and 240-V open-circuit voltage would have a secondary winding of few turns. To achieve the desired reactance, the leakage path gaps would require a large cross-sectional area and a very small spacing, resulting in an uneconomical and difficult-to-manufacture design. Tapping the primary winding permits the designer to assign any value to the secondary turns and induced open-circuit voltage. Third, the line voltage may exceed the required open-circuit voltage. A typical case is a 480-V line and a 240-V open-circuit voltage requirement. The primary winding must be tapped. Fourth, when autotransformers are used for regulating ballasts, the secondary induced voltage accomplishes the regulation. The winding must have assigned to it a percentage of the open-circuit voltage to accomplish the regulation, because the tap voltage changes directly with the line voltage.

The power rating of the autotransformer will be calculated as the product of the rms current and the maximum rms voltage in each winding. The phasor diagram of Figure 4.6 shows that the currents are nearly in phase and will be assumed as such. The tap voltage will be defined as $V_{0t} = kV_0$.

The rating of the primary winding is

$$\begin{aligned} P_p &= I_{p1} V_{0t} + I_{p2} \ (V_0 - V_{0t}) \\ &= I_{p1} k V_0 + I_{p2} V_0 (1 - k). \end{aligned} \tag{4.14}$$

The rating of the secondary winding is

$$P_s = I_n V_2, \tag{4.15}$$

where V_2 is the open-circuit induced voltage and is the maximum that will occur in operation as a leakage-reactance transformer. In a regulating transformer, the secondary winding voltage does rise above the open-circuit value.

The load sets the requirement for the total open-circuit voltage V_{0q}, so that

$$V_2 = V_{0q} - V_{0t} = V_{0q} - kV_0. \tag{4.16}$$

The currents can be found by assuming that the turns are proportional to the open-circuit voltages. Ampere-turn balance requires that

$$\begin{aligned} I_{p1} N_{p1} + I_{p2} N_{p2} &= I_n N_s, \\ I_{p1} k V_0 + I_{p2} (1 - k) V_0 &= I_n (V_{0q} - kV_0). \end{aligned} \tag{4.17}$$

The node equation at the tap point

$$I_{p1} = I_{p2} - I_n, \tag{4.18}$$

so that

$$\left(I_{p2} - I_n\right) kV_0 + I_{p2}(1-k)V_0 = I_n\,(V_{0q} - kV_0),$$

$$I_{p2} = I_n\left(\frac{V_{0q}}{V_0}\right), \tag{4.19}$$

$$I_{p1} = I_n\left(\frac{V_{0q}}{V_0} - 1\right). \tag{4.20}$$

The rating as a function of the tap point is thus

$$P_t = P_p + P_s = I_n\left(\frac{V_{0q}}{V_0} - 1\right) kV_0 + I_n\left(\frac{V_{0q}}{V_0}\right) V_0(1-k)$$

$$+ I_n\left(V_{0q} - kV_0\right),$$

$$P_t = 2I_n\left(V_{0q} - kV_0\right). \tag{4.21}$$

Equation 4.21 shows the power rating as a function of the fixed parameters V_0, V_{0q} and I_n, and the variable k. Two cases can be checked. If the tap point is at the top of the winding, $k = 1$, and the power rating is

$$P_t = 2I_n\,(V_{0q} - V_0) = 2I_nV_2. \tag{4.22}$$

The autotransformer must just transform the series winding power. The "2" designates the rating of the primary and secondary windings separately.

If the tap point is at the bottom, $k = 0$, and the rating is

$$P_t = 2I_nV_{0q}. \tag{4.23}$$

The autotransformer must transform the total power, which is considered as open-circuit voltage times load current. The rating varies linearly from being proportional to the open-circuit secondary winding voltage V_2 alone for $k = 1$, to the full open-circuit voltage V_{0q} for $k = 0$.

4.7 Summary

Leakage-reactance transformers utilize shunts with air gaps to increase the leakage flux between transformer windings to obtain additional series reactance. The transformers are conveniently analyzed using a pi equivalent circuit. The autotransformer type can be analyzed in the same way but imposes an additional constraint between the primary and secondary currents.

5 REGULATING TRANSFORMERS

Regulating transformers utilize the self-saturation of a portion of their magnetic circuits to regulate some parameter of the load without benefit of feedback amplification. They are divided into two types: the constant-voltage transformer and the constant-wattage or regulating lamp ballast transformer. Each type utilizes a capacitor in its secondary circuit, uses decoupling reactance, and operates on similar principles. They are usually built with the leakage-reactance structure described in Chapter 4. In spite of the large number of these regulating transformers that are built annually, very little sound analytical work has been done on their magnetic circuits.*† They are designed on a cut-and-try process and frequently manufactured by adjusting each one before completion. The computer presents a promising means for analyzing and designing regulating transformers, because it can be programmed to include the nonlinear characteristics of the core material. The equations and analytical procedure are developed in this chapter for the regulating transformer. The detailed emphasis is placed upon the ballast.

5.1 Modes of Regulation

As the line voltage is varied above and below its nominal value the regulating transformer maintains the load voltage or lamp power within a narrow range about its nominal value by one of two modes, depending upon the design of the transformer. In the first mode, called impedance control, the transformer increases the impedance in series with the load to absorb the increased line voltage and limit the rise in current. In the second mode, called

*R. N. Basu, "A new approach in the analysis and design of a ferroresonant transformer," *IEEE Trans. on Magnetics,* vol. MAG-3, no. 1, pp. 43–49, March 1967.

†L. A. Finzi and A. Lavi, "The controlled ferroresonant transformer," *AIEE Trans. (Commun. and Electron.)*, vol. 81, pp. 414–418, January 1963.

flux control, the transformer reduces the coupling between the line and load circuit so that the increase of line voltage is not applied to the load. All magnetic-circuit-type regulating transformers utilize one or the other of these modes as their principal means of operation.

None of the regulating transformers achieves perfect regulation; they can be designed to meet specified levels of regulation for given ranges of line voltage. For example, a lamp power range of ± 2 per cent may be required for a line-voltage range of ± 10 per cent for a regulating lamp ballast. If the lamp voltage changes ∓1 per cent, then the lamp current must be regulated to ± 3 per cent, which is an attenuation of the input voltage change by a factor of about three. Generally, the better the regulation the larger, heavier and more expensive is the transformer. This is particularly true in the comparison of two-winding transformers with smaller regulating autotransformers. The regulation of a constant-voltage transformer is supplemented by a bucking winding coupled to the primary. Unlike a ballast which is designed for a particular lamp, a constant-voltage transformer must regulate for both line voltage and load change.

The impedance control transformer depends upon the interaction of the reactance of a reactor with the current in its windings. When the iron is operated at high flux densities an increase of current reduces the reactance and vice versa for a decrease of current. The reactor is connected in series with a capacitor to form the ballasting reactance for a lamp and in parallel with the load for voltage regulation.

The flux control transformer utilizes the property of the structure to shift magnetic flux from a saturated magnetic path to an unsaturated magnetic path. The secondary winding of the transformer is wound on the saturated path. As the total flux is increased by an increase of line voltage, the reluctance of the saturated path increases and the extra flux is forced into the unsaturated path. Hence the change in voltage induced in the secondary winding is attenuated from the full line-voltage change.

5.2 Impedance Control Circuits

These are the simplest circuits and have limited commercial use. However, they are fundamental to the regulating process. The ballast circuit will be treated first, the constant-voltage regulator, later.

The elementary regulating circuit is shown in Figure 5.1a. The circuit consists of a reactor in series with a capacitor; the net reactance of the two elements constitutes the ballasting reactance. For regulation the reactance of the capacitor must exceed the reactance of the reactor; the net reactance magnitude is thus $(X_c - X_l)$. Regulation is accomplished by operating the reactor in the region of magnetic saturation and utilizing its negative change of reactance with current. An increase of line voltage produces an increase of

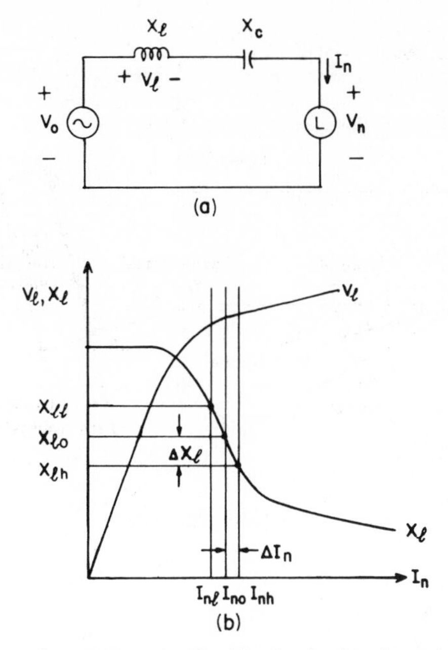

Figure 5.1 Operation of regulating circuit: (a) circuit; (b) dependence of reactance on voltage.

lamp current. The reactance of the reactor decreases, but the net ballasting reactance $(X_c - X_l)$ increases, thereby limiting the change in lamp current to less than the per cent line-voltage change.

The operation is shown more graphically in Figure 5.1b. The magnetization curve of the reactor is shown as voltage V_l versus current I_n. The reactance $X_l = V_l/I_n$ is also sketched. Assume that the nominal lamp current I_{n0} operates the reactor at reactance X_{l0}. A change of lamp current $\pm\Delta I_n$ produces a change of reactance $\mp\Delta X_l$, which appears in the lamp circuit as a change $\pm\Delta X$ and counteracts the change of lamp current. Obviously, if the magnetization curve bends very sharply, the reactance characteristic has a steep slope and the regulation is good. However, there is a practical limit to designing the reactor for such sharp saturation; the lamp current becomes too peaked and may shorten lamp life.

A phasor diagram can only be constructed for a circuit if the electrical quantities are periodic and only for one frequency component at a time. The line voltage will be assumed to be a fundamental voltage source. The lamp will be assumed to be a load at the fundamental frequency and a third-harmonic voltage source. The phasor diagram for the series reactor ballast is shown in Figure 5.2 for the fundamental components of lamp voltage and current, V_{n1} and I_{n1}. The reactor and capacitor losses are represented by a

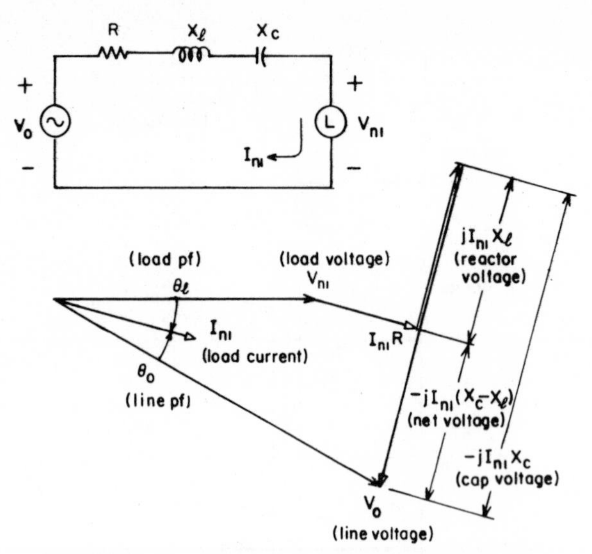

Figure 5.2 Phasor diagram of reactor regulating circuit.

resistance R. The lamp is assumed to have a slightly lagging power factor angle θ_l, of the order of PF = 0.98. The reactor voltage $jI_{n1}X_l$ is shown leading the current I_{n1}, while the capacitor voltage $jI_{n1}X_c$ lags the current. The net power factor for the circuit is leading by the line power factor angle θ_0.

The manner in which the ballast regulates is shown by the diagrams in Figure 5.3 and the reactance characteristic of Figure 5.1. For the low-line condition, the reactance increases to X_{ll} and the diagram shifts as shown in Figure 5.3. For the high-line condition, the reactance decreases to X_{lh} and the diagram shifts as shown in Figure 5.3b.

The interaction of the various parameters in a regulating ballast can best be seen by tabulating their changes with change in line voltage. A set of typical values for a 400-W ballast is shown in Table 5.1. The ballast is based on a nominal open-circuit source voltage of 176 V, lamp quantities of 3.2A , 125 V fundamental component and 400 W power; the ballast consists of a 49-Ω reactor and a 30μF, 88-Ω capacitor. The line voltage is forced to change by ± 10 per cent and the resultant parameters are observed.

Table 5.1 shows that the ± 10 per cent change of source voltage produces a change of reactance of −12, +14 per cent and a current change of ± 6.3 per cent. Because the lamp voltage changes slightly in the opposite direction, the power changes only ± 4.5 per cent. The ballast thus attenuates the per cent line-voltage change by about two. If the reactance X_l were more sensitive to changes of lamp current ΔI_n, the ballast would regulate the lamp power more closely.

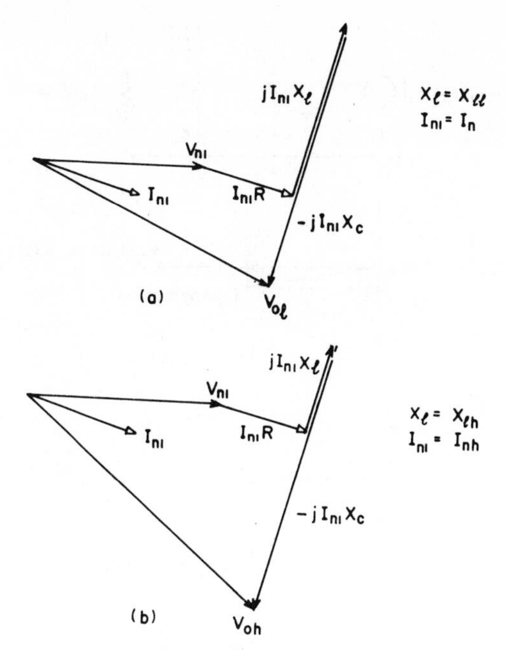

Figure 5.3 Phasor diagrams of regulating circuit of Figure 5.2 for: (a) low line-voltage V_{0l}; (b) high line-voltage V_{0h}.

The impedance control ballast circuit shown in Figures 5.1, 5.2 and 5.3 is essentially a current regulator. It achieves power regulation because the lamp has the typical constant voltage of an arc. The voltage regulating circuit that is analogous to it is shown in Figure 5.4. The circuit regulates load voltage for variations from nominal of a current source. The phasor diagrams of Figure 5.4 for high and low line-voltage conditions show that the load voltage is preserved because the saturation of the reactor allows it to absorb the current variation with small voltage variation. The circuit of Figure 5.4 has small commercial importance because power-level current

Table 5.1 Regulation Operation of Reactor Ballast

Source Volt. V_0	Lamp Current I_n	Lamp Volt. V_{nl}	Lamp Power P_n	Reactor X_l	Net React. $(X_c - X_l)$	Ballast Volt. $I_{nl}(X_c - X_l)$
194V (+10%)	3.4A (+6.3%)	123V (−1.6%)	418W (+4.5%)	43Ω (−12%)	45Ω (+15.4%)	151V
176V (0%)	3.2A (0%)	125V (0%)	400W (0%)	49Ω (0%)	39Ω (0%)	124V
158V (−10%)	3.0A (−6.3%)	127V (+1.6%)	382W (−4.5%)	56Ω (+14%)	32Ω (−18%)	96V

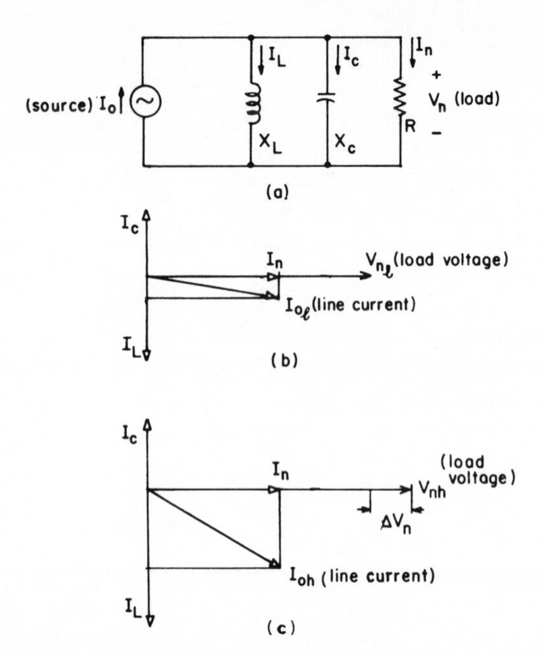

Figure 5.4 Voltage regulating circuit: (a) circuit; (b) phasor diagram for low line current I_{0l}; (c) phasor diagram for high line current I_{0h}.

sources are rarely found. However, the circuit is basic to voltage regulating transformers where the line voltage is converted to a quasi-current source by the series linear reactance element. It is convenient to keep in mind that the constant-wattage ballast and the constant-voltage transformer are very nearly circuit duals.

5.3 Flux Control Circuits

The circuit to be described can be built using two reactors for the two elements in the equivalent circuit, or it can be built using a single magnetic structure that has an equivalent circuit like that shown. The operation of the circuit as a regulating lamp ballast and as a constant-voltage transformer will be described. The impedance control circuit previously described operates by the effect on the reactance of a change of the lamp current. The flux control circuit operates by the effect of the line voltage on the reactance of the shunt element. The elementary form of the circuit is shown in Figure 5.5. The magnetic portion of the circuit has two reactors: a series linear reactor X_l, and a shunt saturable reactor X_m. The regulating action is obtained from the saturable reactor; the series reactor acts to decouple the regulating section from the line and to provide some impedance necessary for the action.

The basic operation of the circuit can be understood by referring to the magnetization curve of X_m shown in Figure 5.5. Assume that the line voltage is at its lowest level and the shunt reactor X_m is operating at voltage V_{ml} and current I_{ml}. The net current, $I_n + I_m$, passes through the series reactor X_l. If the line voltage is raised to its highest level, the shunt reactor current rises to I_{mh}, increasing the lagging reactive current in X_l and absorbing almost the entire voltage increase there. The sharper the knee of the magnetization curve, the better is the regulation. Another way that the action can be described is by viewing the shunt reactor X_m as a voltage limiting device, which maintains nearly constant voltage to the capacitor and lamp circuit. That is, the two reactors form a variable ratio voltage divider for the line voltage. The circuit is also similar to that used for the constant-voltage transformer, where the capacitor and load are connected in parallel with the saturable reactor.

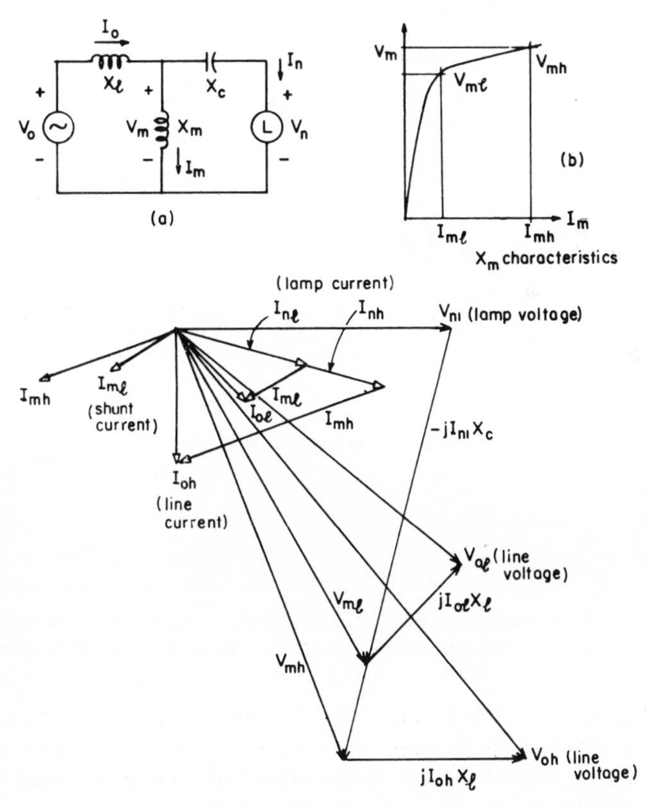

Figure 5.5 Two reactor flux control ballast: (a) circuit; (b) characteristics of shunt reactor X_m; (c) phasor diagram for high and low line-voltage.

The phasor diagram will add to an understanding of the operation, and will illustrate the interaction of the components. The diagram is shown in Figure 5.5 for the upper and lower levels of line voltage. The diagram is drawn for the fundamental-frequency components of voltage and current. As the line voltage is raised from V_{0l} to V_{0h}, the shunt reactor current increases sharply from I_{ml} to I_{mh}. The net current in the series reactor changes sharply in phase from I_{0l} to I_{0h}. The series reactor voltage drop jI_0X_l increases slightly in magnitude and largely in phase to absorb the line-voltage increase. Just as for the impedance-type control circuit, the lamp current must increase slightly from I_{nl} to I_{nh} to accomplish the regulation.

The two-reactor circuit for regulating voltage is shown in Fig. 5.6; it is analogous to the two-reactor circuit for regulating lamp current in Figure 5.5.

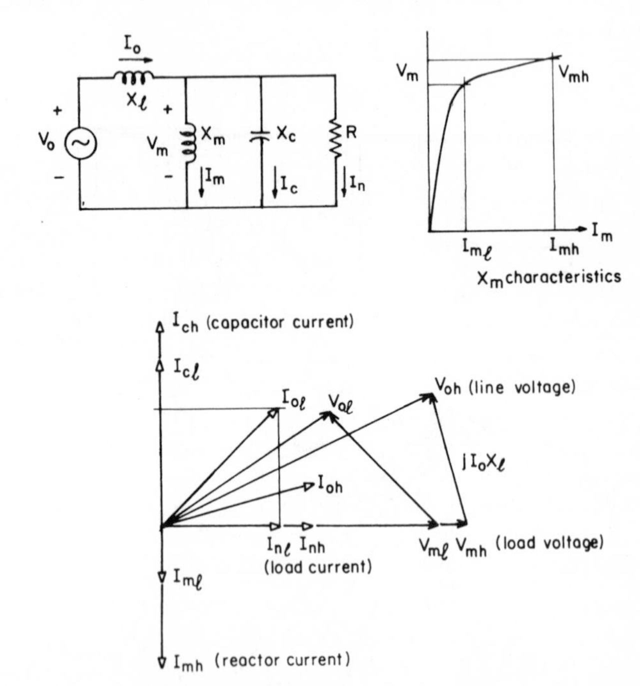

Figure 5.6 Two reactor flux control voltage regulator: (a) equivalent circuit; (b) characteristics of shunt reactor; (c) phasor diagram for high and low line-voltage.

The regulating action is obtained by the large change of current I_m in the shunt reactor for small changes of voltage V_m. The large change of current produces a corresponding large change of voltage drop jI_0X_l across the series reactor and absorbs the change of line voltage. The circuit is usually operated so that the line current I_0 has leading phase with respect to the line voltage

V_0 at low voltage and swings toward zero phase at the highest line voltage of the range. The constant-voltage circuit can be built in two-reactor form but is more commonly built into a leakage-reactance structure where the shunts provide the series reactance and the secondary core provides the saturable shunt reactance.

A convenient equivalent circuit for single magnetic structures used in flux-control ballasts is the pi circuit shown in Figure 5.7. The equivalent circuit

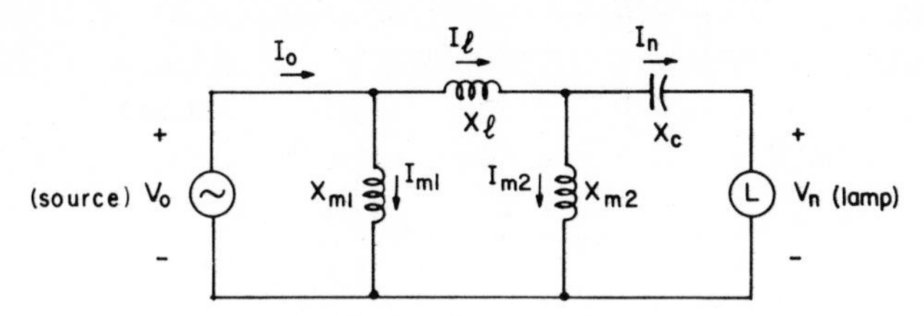

Figure 5.7 Pi equivalent circuit for flux-control ballast.

will be related to the magnetic structures in Section 5.4. The operation of the pi circuit is identical with that of the circuit of Figure 5.5. The effect of the added branch X_{m1} connected across the source is to draw additional current from the source. The element X_{m1} can be used to improve the power factor of the ballast. The pi equivalent circuit of the flux-control ballast shown in Figure 5.7 can be reduced to the form of the impedance control ballast to show similarity and for ease of analysis. Moreover, the elements of the reduced circuit correspond to the portions of the magnetic circuit represented by the pi circuit, as will be shown in Section 5.4. The steps of the reduction are shown in Figure 5.8. It must be emphasized that the final form of the equivalent circuit shown in Figure 5.8c presents to the capacitor and lamp the same open-circuit voltage and internal impedance as the original pi circuit. It does not properly represent the elements of the pi circuit itself.

The first step shown in Figure 5.8a from the circuit of Figure 5.7 shows the reactance X_{m1} absorbed into the source. The circuit is unchanged to the right of *a-b* and sees the correct source and impedance to the left of *a-b*. The second step of Figure 5.8b shows the reactances and source reduced to a Thevenin equivalent circuit to the left of $a' - b'$. The new source voltage V_0' is the open-circuit voltage when one looks back from terminals $a' - b'$ and is given by

$$V_0' = \frac{X_{m2}}{(X_{m2} + X_l)} V_0. \tag{5.1}$$

The new source impedance is that seen looking back from terminals $a' - b'$ with the voltage source V_0 short-circuited. It consists of the reactance X_l

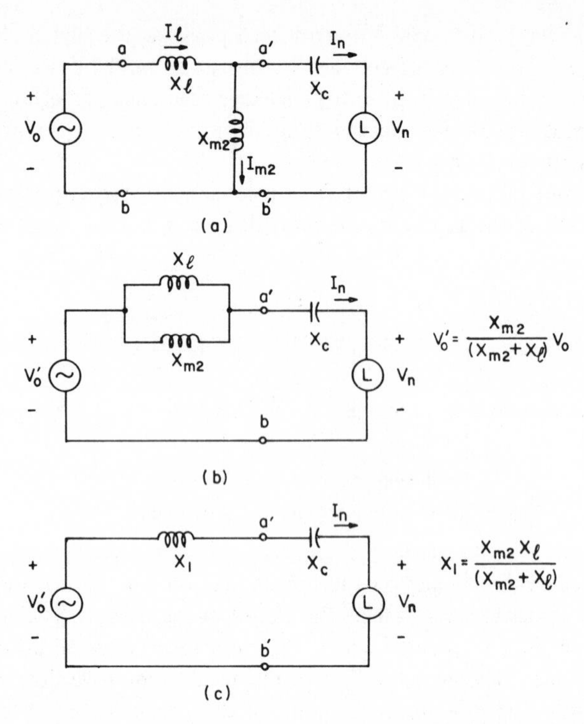

Figure 5.8 Reductions of pi equivalent circuit of Figure 5.6.

and X_{m2} in parallel. The final circuit is shown in Figure 5.8c where the source impedance is

$$X_1 = \frac{X_{m2} X_l}{(X_{m2} + X_l)} . \tag{5.2}$$

We see that the flux control ballast, compared to the impedance control ballast, achieves its regulation both by the effect of the change of saturable reactance X_{m2} on the impedance X_1, and on the source voltage V_0'. Note that the current crossing the $a' - b'$ connection in Figure 5.8c is not the actual line current I_0, nor is the voltage drop $jI_n X_1$, equal to any voltage in the actual pi circuit. The phasor diagram is just like that in Figure 5.2.

A set of typical values for a 400-W ballast as the line voltage is varied is shown in Table 5.2. The values were chosen as close to those for the ballast of Table 5.1 in order to make a comparison possible. The capacitor is 88Ω, the nominal lamp voltage, current, and power are the same, namely, 125V, 3.2A, and 400W. The nominal values of reactance were selected as $X_l = 60\Omega$, $X_{m2} = 250\Omega$ to obtain $X_1 = 49\Omega$, as in the previous example.

The saturable reactance X_{m2} was assumed to vary down to 200Ω at high line-voltage and up to 350Ω at low voltage to produce a ± 6 per cent variation

Table 5.2 Regulation Operation of Pi Circuit Ballast

Eq. Source Volt. V_0'	Source Volt. V_0	Lamp Current I_n	Lamp Volt. V_{n1}	Lamp Power P_n	React. X_{m2}	Equiv. React. X_1	Net React. $(X_c - X_1)$
187V (+6%)	243V (+11.5%)	3.3A (+3.1%)	124V (−0.8%)	410W (+2.5%)	200Ω (−20%)	46Ω (−6%)	42Ω (+7.7%)
176V (0%)	218V (0%)	3.2A (0%)	125V (0%)	400W (0%)	250Ω (0%)	49Ω (0%)	39Ω (0%)
168V (−4.5%)	195V (−10.5%)	3.1A (−3.1%)	126V (+0.8%)	390W (−2.5%)	350Ω (+40%)	52Ω (+6%)	36Ω (−7.7%)

of X_1, which is only one-half that for X_l in the previous example. These variations of X_1 could only account for source voltage variation of + 6 per cent to −4.5 per cent in an impedance control ballast. However, because of the control shown by Equation 5.2, the variation of source voltage is +11.5 per cent to −10.5 per cent. The lamp power is regulated to ±2.5 per cent.

5.4 Magnetic Operation

The reactances shown in the equivalent circuits of the regulating transformers represent portions of magnetic circuits, either as individual reactors or as parts of transformers.

The reactor is the elementary nonlinear component that provides the regulation mechanism of the regulating transformer. As described in Chapter 2, the reactor consists of a laminated steel core with an air gap and winding. The gap reduces the nonlinearity between the instantaneous voltage and current, but reduces the change of regulating reactance for change of current. The magnetic behavior of the reactor is a composite of the behavior of the core material and the gap. The relationship is shown in Figure 5.9 in which the

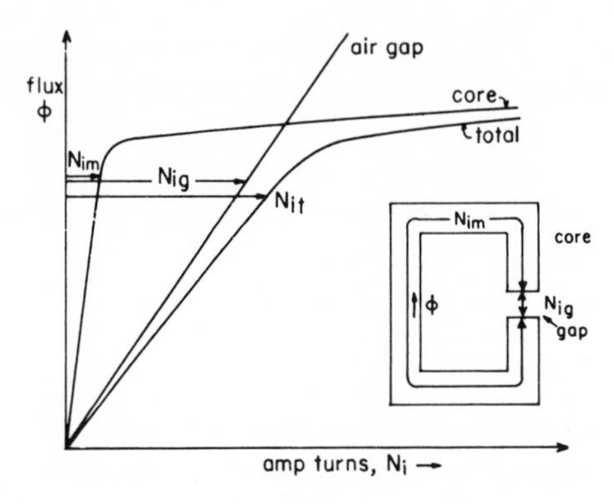

Figure 5.9 Composite magnetization curve for reactor.

ampere-turns Ni_m for the core are added to the ampere-turns Ni_g for the gap at each value of flux ϕ to obtain the composite curve. The curve for the core material is the instantaneous relationship, neglecting hysteresis, corresponding to the normal magnetization curve. The composite curve is less nonlinear than the curve for the core material alone.

Several types of magnetization curves can be measured for a reactor, some physically meaningful and some not. They include curves of rms currents and voltages, rectified (average) currents and voltages, fundamental-frequency component amplitude currents and voltages, and so forth. Two constructions are shown in Figure 5.10 for an inductor driven by a sine wave of current and a

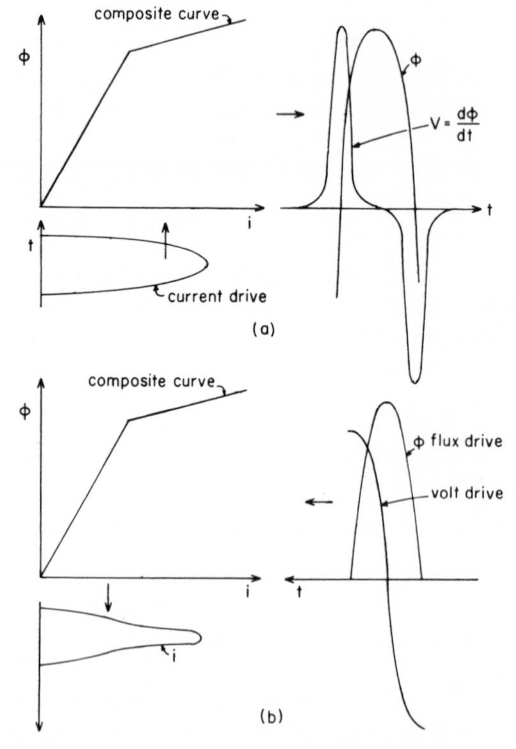

Figure 5.10 Transfer waveforms for reactor: (a) sine wave current, (b) sine wave voltage.

sine wave of voltage. The results are peaked waves of voltage and current, conversely, or a train of odd harmonics.

The circuit description of the reactor is the reactance X_1, which is the ratio of fundamental-frequency voltage to current. For sine-wave current drive, this reactance is calculated by analyzing the voltage waveform for parametric

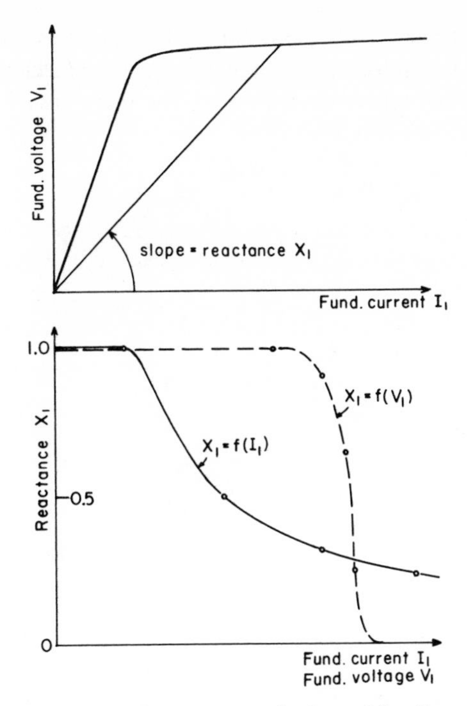

Figure 5.11 Dependence of reactance on fundamental voltage and current.

values of current and plotting the relationship as shown in Figure 5.11. For small harmonic amplitudes, say up to 10 per cent, the rms values of current and voltage will yield the correct values of reactance. Above that level, the reactance measured from rms quantities will be too large with current drive and too small with voltage drive. The reactance is the slope of the line to the operating point on the curve. The dependence of the reactance on the fundamental voltage and current is evident; the sharp change of reactance with voltage is one reason for the better regulation of the flux control ballast.

From Chapter 2, we know that the reactance can be written as the combination of two reactances in parallel, core reactance $X_m = (2\pi f)\,N^2 \mathcal{P}_m$ and gap reactance $X_g = (2\pi f)\,N^2 \mathcal{P}_g$

$$X_1 = (2\pi f)\,N^2 \frac{\mathcal{P}_m\,\mathcal{P}_g}{\mathcal{P}_m + \mathcal{P}_g} = \frac{X_m X_g}{X_m + X_g}. \tag{5.3}$$

The reactance component X_m for the core is a function of the total current or the voltage across the elements. The numerical value of X_1 can be calculated from the core material characteristic of permeability μ_m versus flux density B_m.

A convenient means for calculating the effect of core saturation on the reactance is to use an equivalent air gap made up of the mechanical gap g_0 plus a delta gap Δg which represents the core. The permeance for a physical reactor is

$$\mathcal{P} = \frac{\mu_0 A_g}{g_0 + l_c \left(\frac{\mu_0}{\mu_m}\right) \left(\frac{A_g}{A_m}\right)}, \tag{5.4}$$

where

$$\Delta g = l_c \left(\frac{\mu_0}{\mu_m}\right) \left(\frac{A_g}{A_m}\right).$$

The reactance is thus

$$X_1 = (2\pi f) N^2 \frac{\mu_0 A_g}{g_0 + \Delta g}. \tag{5.5}$$

Saturation of the core reduces the value of permeability μ_m and increases the term Δg which acts as an increase of gap length and a reduction of reactance.

A typical reactor has a mechanical gap $g_0 = 0.040$ in and core length $l_m = 10$ in. The permeability μ_m varies from 1000 μ_0 to 500 μ_0 and $A_g = A_m$. The equivalent gap varies from $g' = 0.040 + 0.010 = 0.050$ in to $g' = 0.040 + 0.020 = 0.060$ in, which produces a reactance change of 20 per cent.

The leakage-reactance transformer such as shown in Figure 5.12 used with a series capacitor in a lamp circuit is a flux control type regulating ballast. When it is used with a shunt capacitor, it acts as a constant-voltage transformer. The regulation for the circuit shown in Figure 5.8 is obtained both by the effect of the regulating reactance X_{m2} on the open-circuit voltage V_0', and on the series reactance X_1. These two operations will be shown to correspond to phenomena in the magnetic circuit of a transformer.

The open-circuit operation of the transformer is described in Figure 5.13 in terms of the secondary core structure. The primary flux ϕ_1 divides into a secondary flux ϕ_2' and a leakage path flux ϕ_l'. The secondary flux links with the secondary winding N_2. The magnetic potential between points x and y is provided by the primary winding ampere-turns $I_0 N_1$, less those required for the primary core path. The equivalent circuit for the open-circuit condition is shown in Figure 5.13b. The voltages correspond to fluxes, the currents to ampere-turns, and the reactances to magnetic paths. The voltage V_0 is that induced by the flux ϕ_1; the voltage V_m' is that induced by ϕ_2'; the voltage $jI_l'x_l$ is that induced by the leakage flux ϕ_l' in a winding of N_2 turns. The current I_l', which equals I_{m2}', produces the ampere-turns $I_l' N_2 = I_{m2}' N_2$ re-

Figure 5.12 Leakage-reactance transformer with bar shunts.

quired for either path between points x and y. The reactance X_{m2} is that of the core path from x to y; the reactance X_l is that of the leakage path all referred to N_2 turns.

As the line voltage is raised, ϕ_1 increases and the secondary core becomes more saturated. Its reactance X_{m2} decreases, reducing the value of V'_m relative to V_0; or, its flux ϕ'_2 decreases relative to ϕ_1, throwing more flux into the leakage paths. The phasor diagram is shown in Figure 5.13c. The flux components ϕ'_2, ϕ'_l, and ϕ_1 are shown with their corresponding voltages, V'_m, jI'_lX_l, and V_0. The voltage V'_m is recognized as the equivalent voltage V'_0 of Figure 5.8.

The effect of load current on the leakage-reactance transformer is shown in Figure 5.14. The load current I_n produces an additional flux component ϕ_n in the secondary core as shown in Figure 5.14a. The ampere-turns I_nN_2 act around the loop covered by the flux ϕ_n. The time phase of the load flux component ϕ_n and the secondary flux ϕ_2 depends upon the phase of the load current I_n with respect to the secondary voltage V''_m. The equivalent circuit for the load-current operation of Figure 5.14a is shown in Figure 5.14b. The source voltage V_0 is zero because the primary flux ϕ_1 is zero. The parallel combination of the reactances X_{m2} and X_l is X_1, which is the reactance of a reactor built like the secondary magnetic circuit including the leakage paths. The current I_n divides into the components I_{nl} and I_{n2} in proportion to the division of ampere-turns I_nN_2 into $I_{nl}N_2$ for the gap and $I_{n2}N_2$ for the core. The impact of an increase of current I_n on the saturation and the reactance X_{m2} depends upon the phase of the flux components ϕ_n and ϕ_2'.

The phasor diagram of Figure 5.14c shows the open-circuit and load components and the resultant values. The load current I_n is assumed to lead the source voltage V_0 by angle θ_0 because of the load-circuit capacitance. The flux ϕ_n is in phase with the current I_n and in the direction shown in Figure 5.14a. The total secondary flux ϕ_2 is thus the sum of ϕ_2' and $-\phi_n$, and is larger than each. This flux induces the total secondary voltage V_m, which is applied to the load circuit. The leakage path flux ϕ_{l2} is the sum of the load flux ϕ_n and the open-circuit leakage flux ϕ_{l2}'. This flux ϕ_{l2} induces the fictitious reactance voltage V_{l2}, which is also $j(I_n + I_{m2}) X_l$. However, only the

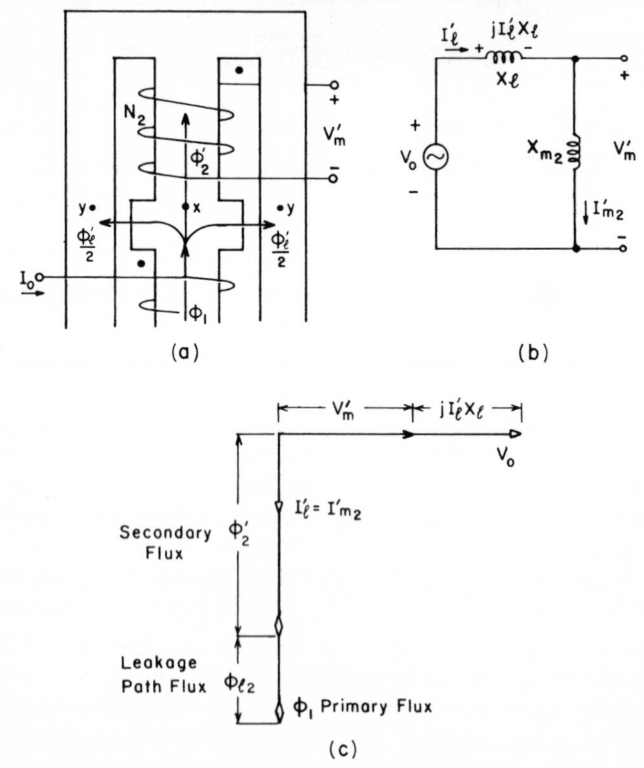

Figure 5.13 Leakage-reactance transformer at open circuit: (a) secondary magnetic circuit; (b) equivalent circuit; (c) phasor diagram.

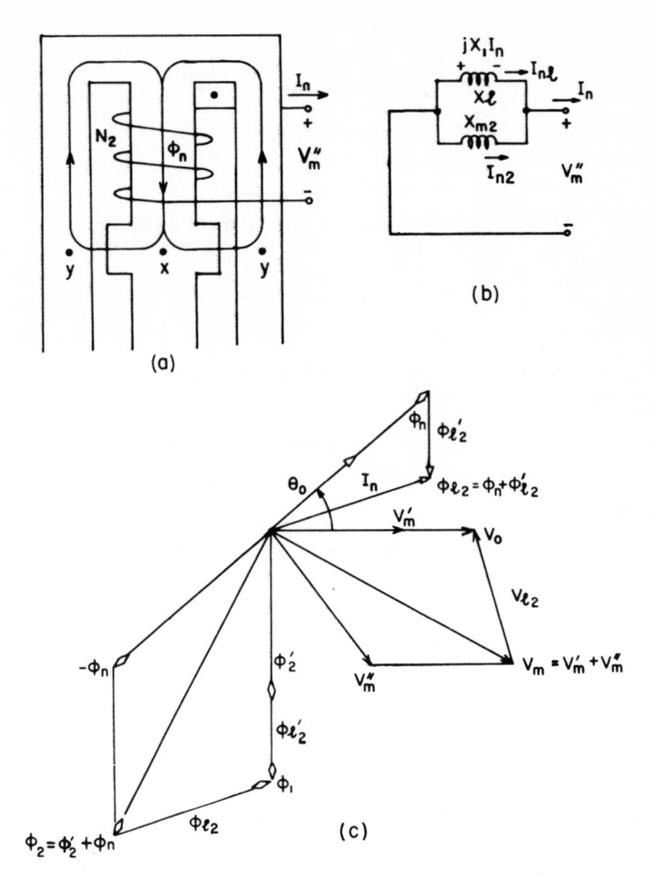

Figure 5.14 Leakage-reactance transformer under load in a regulating circuit: (a) secondary magnetic circuit; (b) equivalent circuit; (c) phasor diagram.

flux ϕ_n induces the fictitious reactance voltage jI_nX_1 of Figure 5.8c. Note that in the first case, the source voltage is V_0 corresponding to flux ϕ_1; in the second case, the source voltage is $V'_n = V'_0$, corresponding to the flux ϕ'_2 of Figure 5.13a. The phasor diagram shows that an increase of the line voltage or the load current will act to increase the total flux ϕ_2, raise the saturation, and both reduce the reactance and shift the flux into the leakage paths.

5.5 Optimization of Regulation

The factors which affect the regulation can be identified by using a simplified model of the regulating circuit. A suitable model is shown in Figure 5.15, where the capacitor is assumed to be the only impedance in the load circuit; the capacitive reactance is usually three times the lamp impedance, so

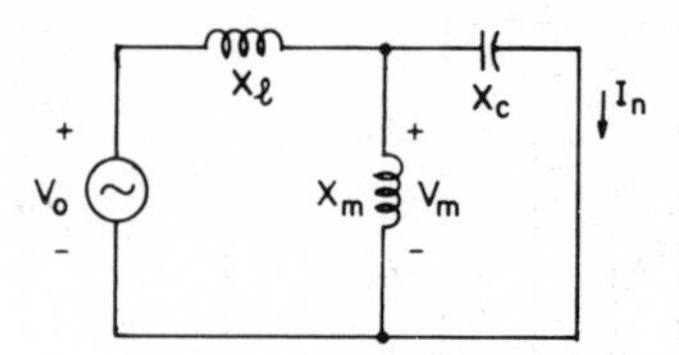

Figure 5.15 Model circuit for calculation of regulation.

the model is reasonable. The model will be used to find the effect of the circuit parameters on the ratio of line-voltage change to load current change. The larger the ratio the better is the regulation.

The load current I_n is given by

$$I_n = \frac{V_0 X_m}{X_c(X_l + X_m) - X_l X_m}. \tag{5.6}$$

The derivative of the line voltage V_0 with respect to I_n, keeping in mind that X_m is not a constant, is

$$\frac{\partial V_0}{\partial I_n} = \frac{-X_c X_l I_n}{X_m^2}\left(\frac{\partial X_m}{\partial I_n}\right) + \frac{X_c X_l}{X_m} + (X_c - X_l). \tag{5.7}$$

Since

$$\frac{\partial X_m}{\partial I_n} = \frac{\partial X_m}{\partial V_m} \cdot \frac{\partial V_m}{\partial I_n} = X_c \frac{\partial X_m}{\partial V_m}, \tag{5.8}$$

the derivative of V_0 in terms of the sensitivity of X_m to voltage V_m becomes

$$\frac{\partial V_0}{\partial I_n} = \frac{-X_c^2 X_l I_n}{X_m^2}\left(\frac{\partial X_m}{\partial V_m}\right) + \frac{X_c X_l}{X_m} + (X_c - X_l). \tag{5.9}$$

For the case of a constant reactance X_m, the derivative is zero and the equation is equal to the ratio of the nominal values V_{00}/I_{n0}

$$\frac{V_{00}}{I_{n0}} = \frac{X_c X_l}{X_m} + (X_c - X_l). \tag{5.10}$$

We can normalize Equation 5.9 with respect to the nominal quantities and obtain

$$\frac{\partial(V_0/V_{00})}{\partial(I_n/I_{n0})} = \frac{-I_{n0}}{V_{00}}\left(\frac{X_c^2 X_l I_n}{X_m^2}\right)\left(\frac{\partial X_m}{\partial V_m}\right) + 1. \tag{5.11}$$

Equation 5.11 shows, as expected, that the constant X_m gives unity normalized regulation, that is, 10 per cent voltage change produces 10 per cent current change.

Equation 5.11 can be further normalized in terms of per unit reactance change per unit voltage change as

$$\frac{\partial(V_0/V_{00})}{\partial(I_n/I_{n0})} = \frac{-I_{n0}}{V_{00}} \left(\frac{X_c X_l}{X_m}\right) \frac{\partial(X_m/X_{m0})}{\partial(V_m/V_{m0})} + 1. \tag{5.12}$$

The negative sign in Equation 5.12 is always cancelled by the negative derivative, the reactance must decrease as the voltage increases. The regulation is improved if X_c or X_l is increased, or if the shunt reactance X_m is reduced. The regulation is also a direct function of the normalized reactance change.

Equation 5.12 is useful for establishing a relationship among the parameters in a computer program when the regulation is stipulated as an input condition to the program.

5.6 *Summary*

Regulating transformers utilize both the change of reactance with current and the flux-shifting properties of a leakage-reactance structure to achieve their results. The operation can be handled in terms of the magnetization characteristic of the core material and the equivalent circuits.

6 NONLINEAR CIRCUIT OPERATION

Two classes of nonlinearity must be considered to insure accurate computer results. First, the electric circuit in which the magnetic device is inserted may have one or more nonlinear elements. A typical example is a gaseous-discharge lamp in a ballast circuit. Second, the magnetic circuit itself may be driven into its saturation region so that both the circuit voltage and current are distorted. An example is a transformer operating in ferroresonance with a capacitor.

6.1 Harmonic Analysis

A useful technique for treating nonlinear circuit problems where one element is nonlinear is to solve the circuit independently for the fundamental-frequency behavior and for each of the significant harmonics. The nonlinear element is handled as a linear element at the fundamental frequency and as a harmonic current or voltage generator at the higher frequencies, which excites the circuit. The solutions for the fundamental and the harmonic cases are then combined to obtain the resultant currents and voltages. The success of the technique depends upon the model which represents the nonlinear element.

A saturating magnetic core is modeled as the usual magnetizing reactance and core loss at the fundamental frequency and as a succession of harmonic voltage generators, each with a characteristic internal reactance. The technique has been described by Blume* for handling overexcited transformers,

*L. F. Blume, *Transformer Engineering,* New York: John Wiley & Sons, 1946, Ch. III.

by Biringer and Slemon* for magnetic frequency multipliers, and by Kusko† for harmonic-suppression filters.

The gaseous-discharge lamp is modeled as a passive impedance at the fundamental frequency and as voltage sources at the harmonic frequencies. The lamp usually operates with nearly sinusoidal current but with distorted voltage. The harmonic analysis of lamp circuits has been described by Zwikker.‡ The lamp circuit will be discussed in detail in this chapter because it represents an application where a computer program must first establish the specifications of the magnetic circuit, then design it.

6.2 Gaseous-Discharge Lamp Model

The simplest operating circuit for a gaseous-discharge lamp is shown in Figure 6.1. The source voltage v_0 has sufficient amplitude to strike the lamp

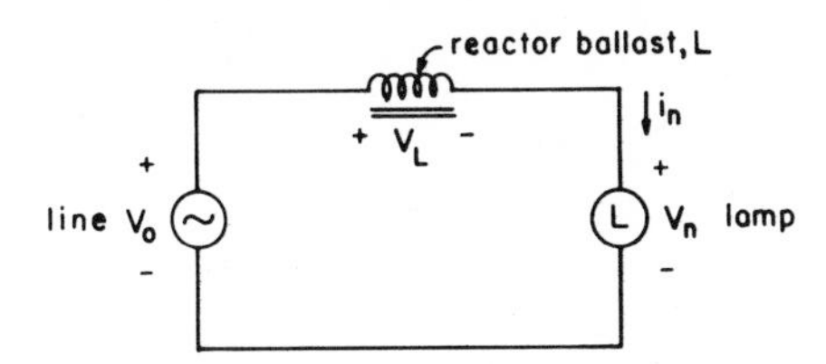

Figure 6.1 Electric circuit for a reactor ballast and lamp.

when the circuit is energized and to maintain the operation. The reactor then absorbs the difference between the source voltage v_0 and the lamp voltage v_n.

Each class of lamp reaches its operating condition in a different way. The low-pressure lamps, such as the flourescent types, reach their steady-state condition in relatively few cycles. The high-pressure lamps, such as the metallic-vapor types, require a warmup period of several minutes before the gas reaches the operating temperature and pressure. The locus of high-pressure lamp starting and warmup with a reactor ballast is shown in Figure 6.2. The lamp strikes at point *a* on the open-circuit voltage, falls to point *b*, then slowly rises to the operating point *c* as the gas pressure builds up to final value. The reactor must be designed to limit the starting lamp current at point *b* while absorbing the full source voltage.

The lamp-voltage waveform is essentially a square wave, while the current satisfies the circuit requirements. The lamp voltage rises each half cycle to a

*P. P. Biringer and G. R. Slemon, "Harmonic analysis of the magnetic frequency tripler," *IEEE Trans. Commun. and Electron.*, vol. 82, pp. 327–332, July 1963.

†A. Kusko, *Power Distribution System,* U. S. Patent No. 2787733, April 2, 1957.

‡C. Zwikker, "The equivalent circuit of a gas discharge lamp," *Philips Tech. Rev.*, vol. 15, no. 6, pp. 161-188, December 1953.

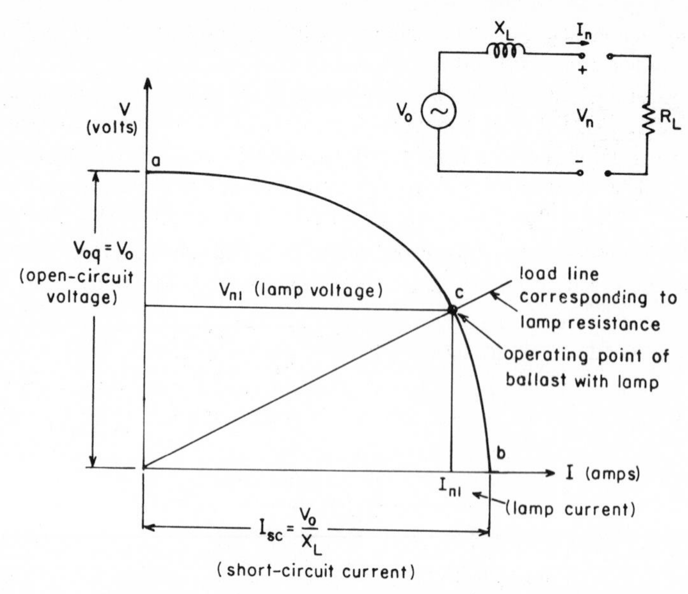

Figure 6.2 Locus of gaseous-discharge lamp operation.

reignition value; then settles to a constant arc-voltage for the duration of the half cycle. The arc voltage is independent of the method of ballasting the lamp and is slightly sensitive to current, an increase of current causes a decrease of arc voltage.

The gaseous-discharge lamp is conveniently modeled as a combination of resistance and inductance at the fundamental frequency and as a voltage source at the third-harmonic frequency. It is not necessary to use more than the third harmonic for good results. The value of resistance and inductance is that which absorbs the rated lamp power and which accounts for the hysteresis in the lamp produced by the temperature variation each half cycle.

The waveforms for the model are shown in Figure 6.3. The third-harmonic lamp voltage v_{n3} is placed in phase with the fundamental to approximate the square-wave lamp voltage. This third-harmonic voltage is then applied to the circuit to find the third-harmonic current i_{n3}. The circuit consists only of the reactor of value $3X_L$ because the source voltage is zero at that frequency. The reactor voltage is then found from the sum of the components v_{L1} and v_{L3}.

A phasor diagram of the fundamental components of the lamp circuit is shown in Figure 6.4. The lamp voltage V_{n1} is multiplied by a factor, typically 1.08, to account for the circuit losses and yields V'_{n1}. The angle θ_1 accounts for the hysteresis in the lamp behavior and is not the power-factor angle

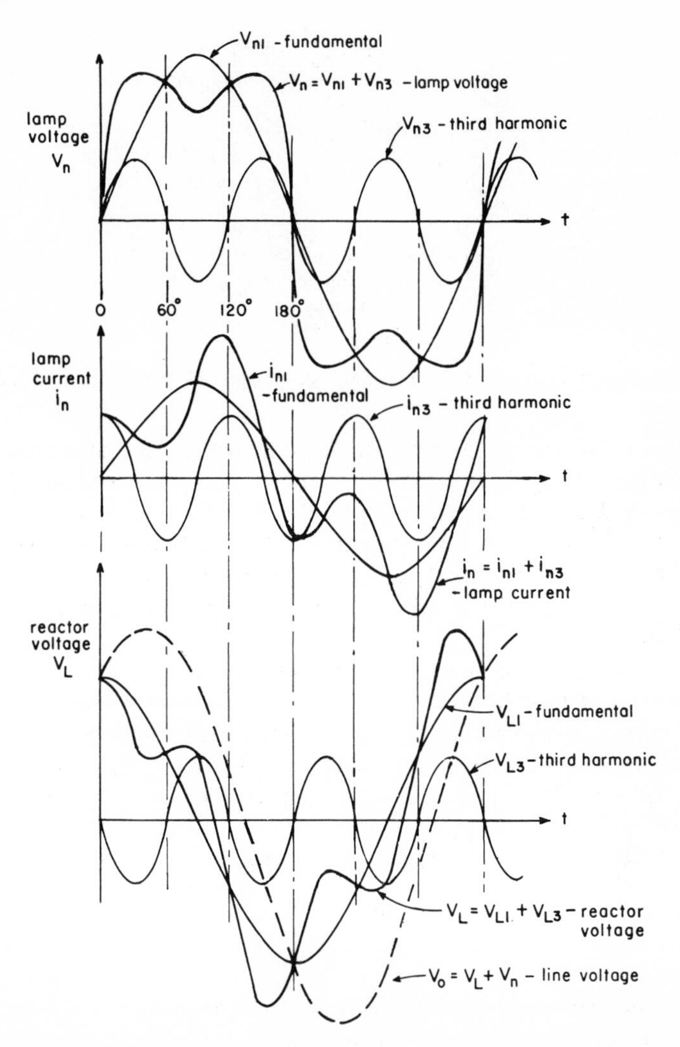

Figure 6.3 Lamp and reactor waveforms using a fundamental and third-harmonic lamp model.

obtained from the rms lamp quantities, that is, $\cos\theta_1 = PFL_1 \neq$ lamp power/lamp voltamperes.

6.3 Design Example

A simple design example for the circuit of Figure 6.1 will be carried out by hand to show how the method of harmonic analysis is used. The electric cir-

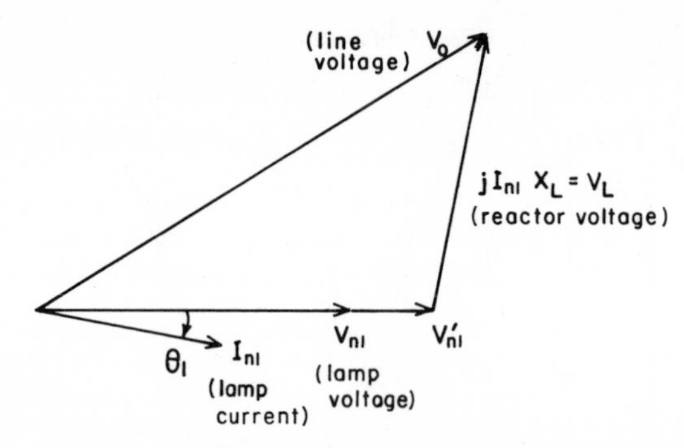

Figure 6.4 Phasor diagram of fundamental components of the circuit of Figure 6.1.

cuit must be solved for the value of choke inductance L that is required to ballast a given lamp for a given line voltage. The circuit will be solved by harmonic analysis using the following assumptions:

1. The waveforms of the lamp current and lamp voltage are adequately described by the fundamental and third-harmonic components.
2. All of the real power to the lamp is carried by the fundamental components.
3. The line voltage is of fundamental frequency only.
4. The ratio of third-harmonic lamp voltage to current is given by the net reactance in series with the lamp at third-harmonic frequency.

The operation of the circuit can be described by five equations as follows: The lamp power in terms of the fundamental current, voltage and power factor

$$P_n = I_{n1} V_{n1}\, \mathrm{PFL}_1. \tag{6.1}$$

The rms lamp current in terms of the current components

$$I_n^2 = I_{n1}^2 + I_{n3}^2. \tag{6.2}$$

The rms lamp voltage in terms of the voltage components

$$V_n^2 = V_{n1}^2 + V_{n3}^2. \tag{6.3}$$

The ratio of third-harmonic components in terms of the circuit reactance at third-harmonic frequency X_3

$$V_{n3}/I_{n3} = X_3. \tag{6.4}$$

The trigonometric equation for the vector diagram of Figure 6.4

$$V_0^2 = (V'_{n1})^2 + (I_{n1}X_L)^2 + 2\,I_{n1}X_L V'_{n1}\sqrt{1-(\mathrm{PFL}_1)^2}. \tag{6.5}$$

Two relationships can be added, namely, $X_3 = 3X_L$ and $V'_{n1} = 1.08\ V'_{n1}$ based on assumed eight per cent loss.

The Equations 6.1 to 6.5 must be solved for the inductance of the reactor. The following variables are usually given for a particular design:

Rated lamp power,	P_n
Rated rms lamp current,	I_n
Rated rms lamp voltage,	V_n
Nominal line voltage,	V_0

Hence, there are five equations with six remaining unknown variables, namely, I_{n1}, I_{n3}, V_{n1}, V_{n3}, X_L, and PFL_1. The value of PFL_1 is known to fall in the range of 0.98 to 1.0, corresponding to a lag angle of zero to 11 degrees. A value of 0.99 can be safely assumed, leaving five unknown variables which can theoretically be found from the five equations.

A solution in closed form for X_L using all of the equations is a formidable algebraic exercise. A simpler iterative approach can be used on either a computer or by hand, starting with the assumption that $I_{n1} \approx I_n$. The value of I_{n3} can be calculated and the resultant I_n calculated. The value of I_{n1} for a second round can be determined by comparing the resultant I_n with the given I_n. Two or three rounds of calculations will force all of the variables to converge to the required accuracy.

As an example consider a 400-W mercury-vapor lamp operating from a 240-V line.

$$P_n = 400 \text{ W}$$
$$I_n = 3.2 \text{ A}$$
$$V_n = 135 \text{ V}$$
$$V_0 = 240 \text{ V}$$

Solve for the inductance of the reactor and the harmonic quantities. Assume $\mathrm{PFL}_1 = 0.99$; $V'_{n1} = 1.08\ V_n$.

Assume for the first round that $I_{n1} = 3.2\text{A}$. Then V_{n1} is found from Equation 6.1 as

$$V_{n1} = \frac{P_n}{I_{n1}\ \mathrm{PFL}_1} = \frac{400}{(3.2)\,(0.99)} = 126.5 \text{ V}.$$

The third-harmonic voltage V_{n3} from Equation 6.3

$$V_{n3} = (V_n^2 - V_{n1}^2)^{1/2} = [(135)^2 - (126.5)^2]^{1/2} = 46.9 \text{ V}.$$

Equation 6.5 in numerical form is

$$(3.2)^2\ X_L^2 + 2(3.2)\,(136.6)\,(0.14)\,X_L - [(240)^2 - (136.6)^2] = 0,$$
$$X_L^2 + 11.95\ X_L - 3900 = 0,$$
$$X_L = 56.8\ \Omega.$$

The third-harmonic current is thus

$$I_{n3} = V_{n3}/3X_L = 46.9/170 = 0.276 \text{ A}.$$

The resultant lamp current is

$$I_n = [(I_{n1})^2 + (I_{n3})^2]^{1/2} = 3.21 \text{ A}.$$

Hence, a second round of calculations is not required and the results to better than one per cent can be listed as

$$\begin{aligned} I_{n1} &= 3.2 \text{ A}, \\ I_{n3} &= 0.276 \text{ A}, \\ V_{n1} &= 126.5 \text{ V}, \\ V'_{n1} &= 136.6 \text{ V}, \\ X_L &= 56.8 \ \Omega, \\ I_{n1}X_L &= 182 \text{ V}, \\ V_{n3} &= 46.9 \text{ V}. \end{aligned}$$

With these numerical values, the reactor can be designed.

6.4 *Summary*

The harmonic analysis method is suitable for a wide range of magnetic circuit problems. Other techniques are more suitable for dc saturable reactors and magnetic amplifiers.* Theoretically, a computer can be programmed to solve for the waveforms of a nonlinear circuit first, then use the waveforms for the design of a magnetic circuit. However, the models of the nonlinear elements, such as the magnetic circuit at high levels of saturation and gaseous-discharge lamps, cannot be formulated that accurately to warrant the extra work and time in the program. It is more profitable to prepare models for the harmonic analysis procedure as a separate problem from the actual design of the magnetic circuit.

*H. F. Storm, *Magnetic Amplifiers,* New York: John Wiley & Sons, 1955.

7 NUMERICAL DESIGN OF REGULATING TRANSFORMERS

The design of regulating transformers which depend upon saturation of portions of their magnetic circuits utilizes all of the material in the previous chapters. The regulating transformer which acts as a gaseous-discharge lamp ballast is of particular interest because it operates with a highly nonlinear load. Its magnetic circuit operation is complex and its parameters are dependent upon its source and load. Numerical design will be treated here, computer design in the next chapter.

7.1 Procedure

The specifications for the design of a regulating transformer ballast are given in terms of the line voltage and its range of variation; the lamp type, power and current; requirements for power factor, starting voltage, outer dimensions and other special factors. The objective is to find the magnetic structure that fulfills the specifications and is economically worthwhile.

The procedure that must be followed in a hand or computer program is the following:

1. Solve the electric circuit problem for the required reactance, waveform, and voltage of the transformer.
2. Design the portion of the magnetic circuit that contributes the series reactance.
3. Design the remainder of the magnetic structure.
4. Calculate the temperature rise, regulation, cost, etc. Modify the input variables and iterate until the solution is obtained.

Step 1 above requires the use of the harmonic techniques described in Chapter 6. Steps 2 and 3 utilize the principles of Chapters 2, 3 and 4. Frequently the design can be reduced to the requirement for designing two reactors just as described in Chapter 3. Finally, the methods of Chapter 5 are used to calculate the regulation in Step 4.

7.2 Numerical Design of Regulating Ballast

The initial design of all regulating ballasts is the same. The difference arises in the structure that is used to realize the required reactances.

The input quantities that determine the electrical behavior of the ballast are the following:

1. Rated lamp power, P_n.
2. Rated lamp current, I_n.
3. Rated lamp voltage, V_n.
4. Nominal line voltage V_0.
5. Limits of line voltage, V_{0h} and V_{0l}; limits of lamp power, P_{nh} and P_{nl}.
6. Required open-circuit voltage, V_{0q}.

The input quantities which determine the physical design of the components are the following:

1. Type of structure.
2. B-H curves and core-loss data.
3. Working flux density at nominal line voltage.
4. Mechanical air-gap lengths.
5. Restricting height, width or depth dimensions.
6. Ratio of capacitive reactance to series reactance, $k_q = X_c/X_1$.

The equivalent circuit that all regulating ballasts reduce to is shown in Figure 7.1 and the phasor diagram is shown in Figure 7.2. For single-reactor

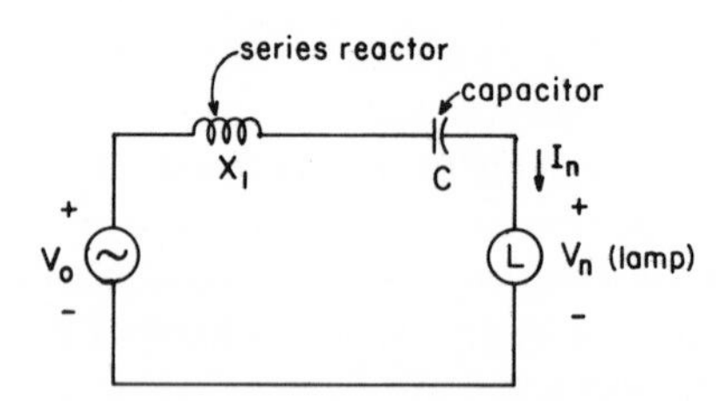

Figure 7.1 Equivalent circuit for regulating ballast circuit.

ballasts, the open-circuit voltage $V_{0q} = V_0$. For transformer and two-reactor ballasts, the open-circuit voltage is $V_{0q} = V_0'$. For single-reactor ballasts, $X_1 = X_l$; for two-reactor ballasts, $X_1 = X_l X_m/(X_l + X_m)$; for transformer ballasts, $X_1 = X_l X_{m2}/(X_l + X_{m2})$.

The electric circuit will be solved by harmonic analysis following the procedure of Chapter 6. The phasor diagram and equivalent circuit of Figure 6.2 are for the fundamental quantities. The reactors are assumed to

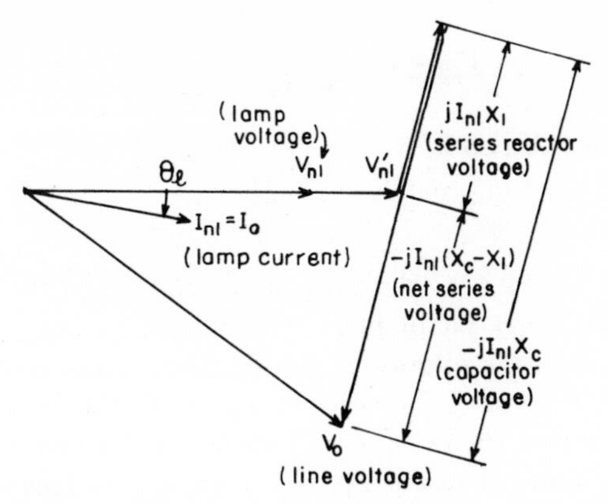

Figure 7.2 Phasor diagram of regulating ballast circuit of Figure 7.1.

be sufficiently linear that they do not generate significant harmonics on their own; only the lamp is a harmonic generator.

The lamp power is given by

$$P_n = I_{n1} V_{n1} \mathrm{PFL}_1. \tag{7.1}$$

The rms lamp current and voltage are given by

$$I_n^2 = I_{n1}^2 + I_{n3}^2, \tag{7.2}$$

$$V_n^2 = V_{n1}^2 + V_{n3}^2. \tag{7.3}$$

The equation for the phasor diagram of Figure 7.2 is

$$[I_{n1}(X_c - X_1)] = V'_{n1}\sqrt{1 - \mathrm{PFL}_1^2} + \sqrt{(V'_{n1})^2(1 - \mathrm{PFL}_1^2) - (V'_{n1})^2 + (V_{0q})^2}, \tag{7.4}$$

where $V'_{n1} = 1.08 V_{n1}$ to account for circuit losses.

The individual reactance components are defined by a parameter selected by the designer, for example, $k_q = X_c/X_1$. They are

$$X_1 = \frac{[X_c - X_1]}{(k_q - 1)}, \tag{7.5}$$

$$X_c = k_q X_1 = \frac{k_q}{k_q - 1}[X_c - X_1]. \tag{7.6}$$

The harmonic voltages on the components are

$$V_{13} = V_{n3}\left[\frac{3X_1}{3X_1 - X_c/3}\right], \tag{7.7}$$

$$V_{c3} = V_{n3}\left[\frac{X_c/3}{3X_1 - X_c/3}\right]. \tag{7.8}$$

The third-harmonic current is

$$I_{n3} = \frac{V_{13}}{3X_1}. \tag{7.9}$$

Equations 7.1 to 7.9 are solved for the parameters of the ballast by first assuming a lamp power factor $PFL_1 = 0.99$, then iterating on the current component I_{n1}, until the solution is found. The current component is first set $I_{n1} = I_n$ and the voltage component V_{n1} is found from Equation 7.1. Equation 7.4 is solved for $[X_c - X_1]$, from which the rest of the equations can be solved including the value I_{n3} of Equation 7.9. This value is inserted in Equation 7.2 to find I_n and to establish a second round value for I_n as follows:

$$(I_{n1})_{\text{corrected}} = (I_{n1})_{\text{original}} \times \frac{(I_n)_{\text{rated}}}{(I_n)\ \text{Eq. 7.2}}. \tag{7.10}$$

As an example use the same requirements as for Section 6.3 with the additional requirement that $k_q = 2$.

$$P_n = 400\ \text{W}$$

$$I_n = 3.2\ \text{A}$$

$$V_n = 135\ \text{V}$$

$$V_{0q} = 240\ \text{V}$$

Assume that $V'_{n1} = 1.08\ V_n$ and $PFL_1 = 0.99$. Assume for the first round $I_{n1} = 3.2$ A.

The fundamental lamp voltage from Equation 7.1 is

$$V_{n1} = \frac{400}{3.2 \times 0.99} = 126.5\ \text{V},$$

$$V'_{n1} = 1.08 \times 126.5 = 136.6\ \text{V}.$$

The net reactance voltage from Equation 7.4 is

$$[I_{n1}(X_c - X_1)] = (136.6 \times 0.14) + \sqrt{(136.6 \times 0.14)^2 - (136.6)^2 + (240)^2}$$

$$= 19.1 + 199 = 218\ \text{V},$$

$$[X_c - X_1] = \frac{218}{3.2} = 68.3\ \Omega.$$

The components are

$$X_1 = 68.3\ \Omega,$$

$$X_c = 136.6\ \Omega.$$

The harmonic voltages are

$$V_{n3} = \sqrt{(135)^2 - (126.5)^2} = 47 \text{ V},$$

$$V_{13} = 47 \frac{205}{(205 - 45.5)} = 60.5 \text{ V},$$

$$V_{c3} = 47 \frac{45 - 5}{(205 - 45.5)} = 13.4 \text{ V}.$$

The third-harmonic current is

$$I_{n3} = \frac{60.5}{205} = 0.295 \text{ A}.$$

The resultant lamp current is

$$I_n = \sqrt{(3.2)^2 + (0.295)^2} = 3.22 \text{ A}.$$

The value is within one per cent so no further calculation is required.

The value of the capacitor is calculated from Equation 7.6 and the line frequency f as

$$C = \frac{10^6}{2\pi f X_c} \mu\text{F}. \tag{7.11}$$

The rms capacitor voltage is given by

$$V_c = [V_{c1}^2 + V_{c3}^2]^{1/2}. \tag{7.12}$$

For the example

$$C = \frac{10^6}{2\pi \times 60 \times 136.6} = 19.5 \ \mu\text{F},$$

$$V_c = [(3.2 \times 136.6)^2 + (13.4)^2] = 437 \text{ V}.$$

The designer can use the capacitance C as an input quantity instead of the reactance ratio k_q to determine the relative values of X_c and X_1.

7.3 Specification of Magnetic Circuits

The conversion of the calculated value of X_1, V_{11}, V_{13}, and I_n into a specification depends upon the magnetic component used. For the single reactor, the specification for reactance is merely

$$X_1 = X_1. \tag{7.13}$$

The rms reactor current is

$$I_1 = I_n. \tag{7.14}$$

The equivalent fundamental-frequency rms sine-wave voltage to which the reactor should be designed is given by

$$V_1' = I_{n1}X_1 + 1/3\ V_{13}. \tag{7.15}$$

The reactor can now be designed as in Chapter 3. The selection of the flux density will determine the regulation.

The two-reactor ballast of Figure 5.5 would only be used where the line voltage was higher than the required lamp open-circuit voltage V_{0q} but still not warrant a transformer. Hence, we can define a coupling coefficient $k_c = V_{0q}/V_0$ as

$$k_c = \frac{X_m}{(X_l + X_m)}. \tag{7.16}$$

The reactances are also related to X_1, which we determined from the electric circuit by

$$X_1 = \frac{X_l X_m}{(X_l + X_m)}. \tag{7.17}$$

The two equations can be solved for the reactances to yield

$$X_m = \frac{X_1}{(1 - k_c)}, \tag{7.18}$$

$$X_l = \frac{X_1}{k_c}. \tag{7.19}$$

The voltage V_m and current I_m of the shunt reactor are found from the phasor diagram of Figure 5.5. The fundamental voltage is given by

$$V_{m1} = \left[V_{n1}^2 + (I_{n1}X_c)^2 - 2\,(V_{n1}'X_c)\sqrt{1 - \text{PFL}_1^2}\,\right]^{1/2}. \tag{7.20}$$

The third-harmonic component is V_{13} from Equation 7.7. The equivalent rms voltage for design is thus

$$V_m' = V_{m1} + 1/3\ V_{13}. \tag{7.21}$$

The currents are

$$I_{m1} = \frac{V_{m1}}{X_m}, \tag{7.22}$$

$$I_{m3} = \frac{V_{13}}{3X_m}, \tag{7.23}$$

$$I_m = [I_{m1}^2 + I_{m3}^2]^{1/2}.$$

The series reactor current is found from the real and reactive power. The active power is

$$P_1 = I_{n1} V'_{n1} \, (\mathrm{PFL}_1). \tag{7.24}$$

The reactive power is

$$P_q = I_{n1} V_{n1} \, [1 - (\mathrm{PFL}_1^2)]^{1/2} + I_{m1}^2 X_m - I_{n1}^2 X_c. \tag{7.25}$$

The fundamental current I_{01} in the series reactor is

$$I_{01} = \frac{[P_1^2 + P_q^2]^{1/2}}{V_{m1}}. \tag{7.26}$$

The third-harmonic current is

$$I_{03} = \frac{V_{13}}{3X_l}. \tag{7.27}$$

The rms series reactor current is

$$I_0 = [I_{01}^2 + I_{03}^2]^{1/2}. \tag{7.28}$$

The equivalent rms design voltage for the series reactor is

$$V_1' = [(I_{01} X_l) + {}^1/_3 \, V_{13}]. \tag{7.29}$$

The two reactors are now specified. The flux density selected for the shunt reactor sets the regulation.

As an example, calculate the specifications for a two-reactor regulating ballast for the conditions of Section 7.2. The line voltage is $V_0 = 277$ V.

$$k_c = \frac{240}{277} = 0.865,$$

$$X_m = \frac{68.3}{(1 - 0.865)} = 505 \ \Omega,$$

$$X_l = \frac{68.3}{0.865} = 79.0 \ \Omega,$$

$$V_{m1} = \Big[(136.6)^2 + (3.18 \times 136.6)^2 - 2\,(136.6)\,(3.18 \times 136.6)\sqrt{1 - (0.99)^2}\Big]^{1/2} = 438 \text{ V},$$

$$V_m = 438 + 1/3\,(60.5) = 458.2 \text{ V},$$

$$I_{m1} = \frac{438}{505} = 0.87 \text{ A},$$

$$I_{m3} = \frac{60.5}{3 \times 505} = 0.04 \text{ A},$$

$$I_m = 0.87 \text{ A},$$

$$P = (3.18)\,(136.6)\,(0.99) = 432 \text{ W},$$

$$P_q = (3.18)\,(136.6)\,(0.14) + (0.87)^2\,(505) - (3.18)^2\,(136.6)$$
$$= 60.6 + 382 - 1380 = -937 \text{ var},$$

$$I_{01} = \frac{[(432)^2 + (937)^2]^{1/2}}{438} = 2.36 \text{ A},$$

$$I_{03} = \frac{60.5}{3 \times 79} = 0.26 \text{ A},$$

$$I_0 = 2.38 \text{ A},$$

$$V_1' = (2.36 \times 79) + 1/3\,(60.5) = 206.7 \text{ V}.$$

The specification of the leakage-reactance transformer is similar to that for the two reactors. The transformer provides two additional elements that are under the designer's control; they are the turns ratio N_2/N_1 and the shunt reactance X_{m1} of Figure 5.6. The two elements are used to control the regulation of the ballast since the coupling coefficient k_c can be chosen independently of the voltages, and the input power factor can be set. Generally the lower k_c is made, that is, the lower the coupling, the better is the regulation and the larger and heavier is the ballast.

The coefficient k_c is selected as typically 0.85. The parameters are given by

$$\frac{N_2}{N_1} = \frac{1}{k_c}\frac{V_{0q}}{V_0}, \tag{7.30}$$

$$X_{m2} = \frac{X_1}{(1 - k_c)}, \tag{7.31}$$

$$X_l = \frac{X_1}{k_c}. \tag{7.32}$$

The voltage V_m' of Equation 7.21 is the secondary winding voltage at full load and determines the relationship between turns N_2 and flux density. The currents I_m and I_0 for the shunt and series elements are fictitious currents for the transformer. They represent the ampere-turns $I_m N_2$ for the secondary core and $I_0 N_2$ for the leakage path. The secondary winding current is the lamp current I_n. The voltage V_l' of Equation 7.29 is the voltage that would be induced in a winding of N_2 turns carrying all of the leakage path flux. It determines the necessary cross section for the leakage path.

The value of the shunt reactance X_{m1} of Figure 5.6 can be set to obtain a desired input power factor. Let that power factor be PFO. Then, the value of X_m must satisfy

$$\frac{P_{q1}}{P_1} = \frac{[1 - (\text{PFO})^2]^{1/2}}{(\text{PFO})} = \frac{I_{n1} V_{n1} [1 - (\text{PFL})^2]^{1/2} + I_{m1}^2 X_{m2} - I_{n1}^2 X_c}{I_{n1} V_m' \, (\text{PFL})}$$

$$+ I_l^2 X_l + \frac{(V_0 N_2 / N_1)^2}{X_{m1}} \tag{7.33}$$

The fundamental primary winding current is

$$I_{01} = \frac{[P_1^2 + P_{q1}^2]^{1/2}}{V_0}. \tag{7.34}$$

The third-harmonic primary winding current is approximately

$$I_{03} = \frac{V_{13}}{3X_l} \left(\frac{N_2}{N_1} \right). \tag{7.35}$$

The primary winding current is thus

$$I_0 = [I_{01}^2 + I_{03}^2]^{1/2}. \tag{7.36}$$

As an example, calculate the additional specifications to convert the two-reactor set to a transformer for $V_0 = 120$ V and an input power factor PFO = −0.9. Assume $k_c = 0.865$ as before for convenience. Then

$$\frac{N_2}{N_1} = \frac{1}{0.865} \frac{240}{120} = 2.31,$$

$$X_{m2} = 505 \ \Omega,$$

$$X_l = 79 \ \Omega,$$

$$V_m' = 458.2 \text{ V},$$

$$P_{q1} = -937 + (2.38)^2 \, 79 + \frac{(120 \times 2.31)^2}{X_{m1}}$$

$$= -489 + 7.7 \times 10^4 / X_{m1},$$

$$P_1 = 432 \text{ W},$$

$$\frac{P_{q1}}{P_1} = \frac{0.436}{-0.9} = \frac{7.7 \times 10^4 / X_{m1} - 489}{432} = \frac{178}{X_{m1}} - 1.13,$$

$$X_{m1} = \frac{178}{0.645} = 276 \ \Omega.$$

The primary winding currents are

$$I_{01} = \frac{432}{0.9 \times 120} = 4.0 \text{ A},$$

$$I_{03} = 0.26 \times 2.31 = 0.6 \text{ A},$$

$$I_0 = [(4.0)^2 + (0.6)^2]^{1/2} = 4.04 \text{ A}.$$

The leakage-reactance transformer can now be designed.

7.4 Calculation of Regulation

The function of a regulating ballast is to regulate lamp power against changes of line voltage. To handle the function in a design procedure, the regulation must be used as an input quantity or calculated afterwards. Since the calculation of the regulation requires the handling of both the lamp and the magnetic circuit as nonlinear elements, we must make simplifying assumptions and not expect a perfect answer. However, we must be able to differentiate between design choices to achieve required regulation without having to build the ballast first. A technique of calculation will be described which can be carried out by hand or computer. Only the flux control ballast will be treated.

As described in Chapter 5, the regulation is accomplished by the change in saturation with line voltage of the magnetic material of the elements represented as X_m in the two-reactor ballast, and X_{m2} in the transformer ballast. The change in saturation changes the reactance in a direction to maintain constant lamp power.

The easiest way to calculate the regulation is to assume values of lamp current above and below the nominal value and calculate back to the corresponding values of line voltage. The lamp power can be assumed to follow $I_{n1} V_{n1}$ (PFL_1), where I_{n1} is varied but V_{n1} and (PFL_1) remain fixed.

Information on the magnetic circuit or the material is required in the form of a B-H curve or a magnetization curve for the elements X_m or X_{m2}. The calculation of regulation is best illustrated by using the parameters of the example of Section 7.2 for the two-reactor ballast. The magnetization curve for the reactor X_m is given by Figure 7.3.

Assume lamp currents ± 2 per cent from the nominal of 3.18 A and calculate the corresponding line voltages.

$$I_{nh} = 1.02 \times 3.18 = 3.24 \text{ A},$$

$$I_{nl} = 0.98 \times 3.18 = 3.12 \text{ A}.$$

The values of V_{m1} are calculated from Equation 7.20

$$V_{m1h} = [(136.6)^2 + (3.24 \times 136.6)^2 - 2(136.6)(3.24 \times 136.6)(0.14)]^{1/2}$$

$$= 443 \text{ V},$$

$$V_{m1l} = [(136.6)^2 + (3.12 \times 136.6)^2 - 2(136.6)(3.12 \times 136.6)(0.14)]^{1/2}$$

$$= 426 \text{V}.$$

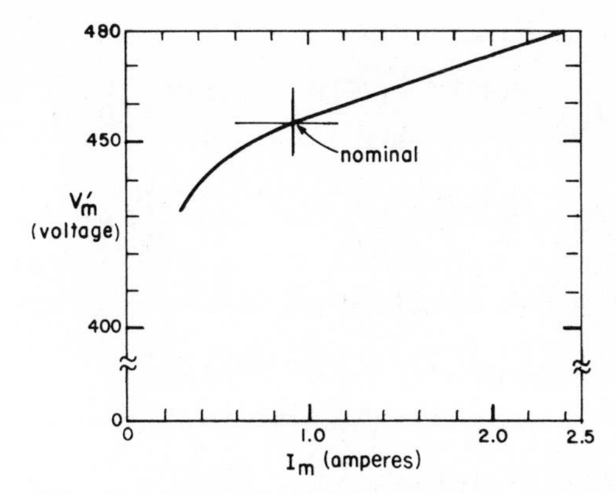

Figure 7.3 Magnetization curve of reactor X_m.

Using Equation 7.21

$$V'_{m1h} = 443 + 1/3(60.5) = 463 \text{ V},$$

$$V'_{m1l} = 426 + 1/3(60.5) = 446 \text{ V}.$$

From Figure 7.3, the magnetization currents are

$$I_{mh} = 1.39 \text{ A},$$

$$I_{ml} = 0.58 \text{ A}.$$

Assuming that $I_{m1} = I_m$, the reactances are

$$X_{mh} = \frac{443}{1.39} = 319\ \Omega,$$

$$X_{ml} = \frac{426}{0.58} = 735\ \Omega.$$

The current in the series reactor is calculated using Equations 7.24, 7.25, and 7.26

$$P'_{nh} = 1.02 \times 432 = 441 \text{ W},$$

$$P'_{nl} = 0.98 \times 432 = 423 \text{ W},$$

$$P_{q1h} = (3.24)(136.6)(0.14) + (1.39)^2 (319) - (3.24)^2 (136.6)$$

$$= 62 + 616 - 1430 = -752 \text{ var},$$

$$P_{q1l} = (3.12)(136.6)(0.14) + (0.58)^2 (735) - (3.12)^2 (136.6)$$

$$= 60 + 247 - 1330 = -1023 \text{ var}.$$

The fundamental series reactor currents are

$$I_{01h} = \frac{[(441)^2 + (752)^2]^{1/2}}{443} = \frac{870}{443} = 1.97 \text{ A},$$

$$I_{01l} = \frac{[(423)^2 + (1023)^2]^{1/2}}{426} = \frac{1110}{426} = 2.60 \text{ A}.$$

The reactive power in the series reactor is

$$(I_{01h})^2 X_l = (1.97)^2 (79) = 307 \text{ var},$$

$$(I_{01l})^2 X_l = (2.60)^2 (79) = 535 \text{ var}.$$

The line voltamperes are given by

$$(VA)_h = [(441)^2 + (752 - 307)^2]^{1/2} = 628 \text{ VA},$$

$$(VA)_l = [(423)^2 + (1023 - 535)^2]^{1/2} = 638 \text{ VA}.$$

The line voltages are

$$V_{0h} = \frac{628}{1.97} = 319 \text{ V},$$

$$V_{0l} = \frac{638}{2.60} = 245 \text{ V}.$$

The calculations show that a change of line voltage from 245 to 319 V around the nominal 277 V, a −11 to +15 per cent change, will change the lamp power by ± 2 per cent. The actual lamp power change will probably be less because the change of lamp voltage tends to counteract the change of lamp current. If the regulation is inadequate, then the circuit must be redesigned to operate the reactor X_m at a higher flux density, or the value of k_q must be decreased.

7.5 Summary

The procedure to be followed by a computer design program must usually be checked by hand calculation both in the program writing and debugging phases. The results of the hand calculation alert the design engineer to inconsistencies in the program and possible difficulties of obtaining convergence in an iterative process loop. In addition, the results provide numerical values for parameters that can be checked with the computer readout. The computer, it must be remembered, only does the hand calculations, but does them faster and more times than the design engineer.

8 COMPUTER DESIGN OF REGULATING TRANSFORMERS

Regulating transformers of either the constant-voltage or constant-wattage type have four elements in common. They utilize a capacitor on the secondary side to interact with the reactance of the transformer; they depend upon magnetic saturation for regulation; they decouple the load from the source with inductive reactance; and they usually employ some form of artificial waveform control. In the constant-voltage transformer, the waveform is modified from an inherent square wave to a sine wave. In the constant-wattage transformer, the waveform of the open-circuit voltage applied to the lamp is modified by adding a pulse to the top. The handling of all of these items in a design is suitable for a computer procedure.* An example of the constant-wattage type will be treated in this chapter.

8.1 Pulse Generation

A voltage pulse can be generated in a winding on a magnetic core by shifting flux into the core for a short period of time each half cycle. The shifting can be done by using cores with air gaps, which do not saturate normally, and closed cores, which do saturate after absorbing a specific volt-time area. A structure for generating a pulse in the secondary winding N_s is shown in Figure 8.1a. The structure consists of a primary winding N_p, connected to a source v_0 of sine-wave line voltage, and enclosing two magnetic cores. The primary magnetic core has an air gap and does not saturate. The secondary magnetic core is solid and relatively narrow so that it would saturate if it tried to support the line voltage without the primary core.

*T. Wroblewski and A. Kusko, "Computer design of the constant wattage choke-capacitor ballast," *National Technical Conference of the Illuminating Engineering Society, Aug. 21 to 26, 1966*, no. 20, pp. 1–4.

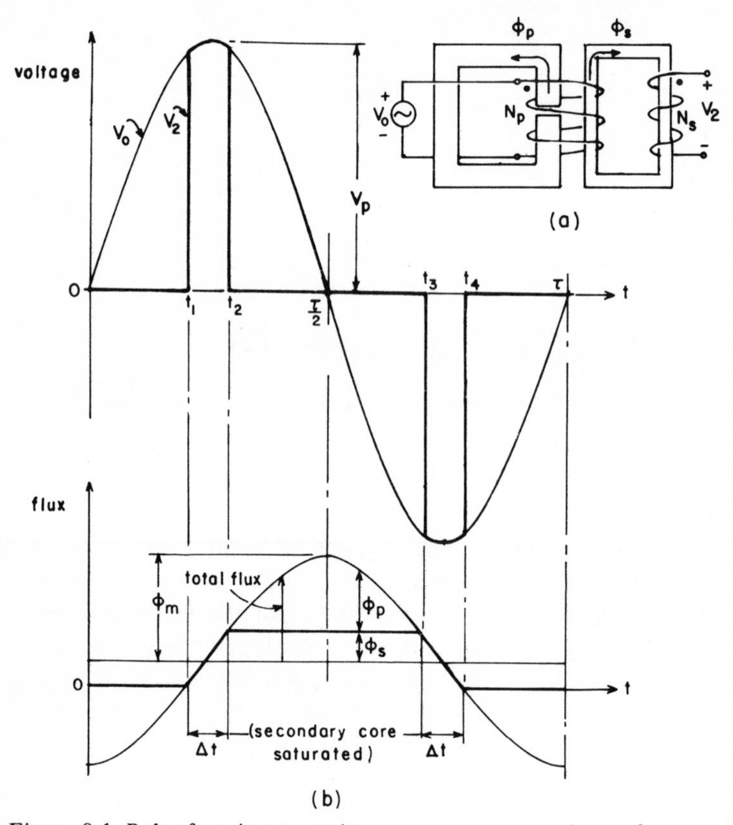

Figure 8.1 Pulse-forming secondary core structure and waveforms.

The circuit operates in two modes: when the secondary core is not saturated, the coupling between the primary and secondary windings is close and the full transformed primary voltage appears on the secondary terminals; when the secondary core is saturated, the coupling is practically zero and no primary voltage appears on the secondary terminals.

The flux and voltage patterns for the pulse-forming structure are shown in Figure 8.1b. The line voltage v_0 impressed on the primary winding demands that the total flux linking the winding must always follow a sinusoid of amplitude ϕ_m, where for the rms value V_0

$$\phi_m = \frac{V_0}{4.44\, N_p f}. \tag{8.1}$$

The primary winding does not care which core carries the flux as long as the total meets the requirement. At any time that the total flux exceeds the saturation flux level of the secondary core, the core will be saturated and the excess flux will be carried by the primary core. At any time that the total flux is less than the saturation level of the secondary core, all of the flux will be carried by the secondary core because the permeance is higher than the primary core.

The pattern can be followed in Figure 8.1b. At $t = 0$, the total flux is negative maximum, $-\phi_m$, and the secondary core is saturated. This condition prevails until $t = t_1$, when the secondary core becomes unsaturated and carries all of the total flux linking the primary winding. At $t = t_2$, the total flux reaches the positive saturation level of the secondary core, the core saturates and the additional flux until $t = t_3$ is carried by the primary core.

During the times 0 to t_1, t_2 to t_3, . . ., the secondary voltage is zero because the flux is constant; during the times t_1 to t_2, t_3 to t_4, . . ., the secondary voltage is $v_0(N_s/N_p)$, because the coupling is close. The pulse voltage can be made as high as required by adjusting the turns ratio N_s/N_p. The width of the pulse is governed by the volt-time area. For a narrow pulse, it is given by

$$V_p(\Delta t) \approx 2N_s\phi_{s,\ \mathrm{sat.}}, \qquad (8.2)$$

where

$$\phi_{s,\ \mathrm{sat.}}$$

is the product of the secondary core area and the saturation flux density. The pulse width is essentially controlled by the core area since the turns are set by the pulse height and the flux density is fixed by the material.

A similar magnetic structure to that of Figure 8.1a can be used to supply a voltage pulse on top of a sine wave.. The new structure is shown in Figure 8.2a. The secondary core has a larger cross section than before and is now equipped with a bridged or splined air gap. The cross section of the bridge is arbitrarily made the same as the secondary core in Figure 8.1a. The bridge acts just like the secondary core of Figure 8.1a. When it is unsaturated, it provides close coupling between the primary and secondary windings. However, when the bridge is saturated, the secondary core acts like a gapped core and the total flux splits between the primary and secondary cores in accordance with their permeances. The secondary winding now has a voltage during the intervals when the bridge is saturated.

The waveforms of flux and voltage are shown in Figure 8.2b. The sine-wave part of the secondary voltage has a peak value V_2' and the pulse part a peak value V_p. As before, the width of the pulse is controlled by the cross-sectional area of the bridge. The amplitude of the sine-wave part is controlled by the relative permeances of the two cores. If the permeance of the primary core is $\mathcal{P}_p$ and the series core $\mathcal{P}_s$ when the bridge is saturated, the value V_2' is given by

$$V_2' = \sqrt{2}\, V_0 \frac{N_s}{N_p} \frac{\mathcal{P}_s}{\mathcal{P}_p + \mathcal{P}_s}. \qquad (8.3)$$

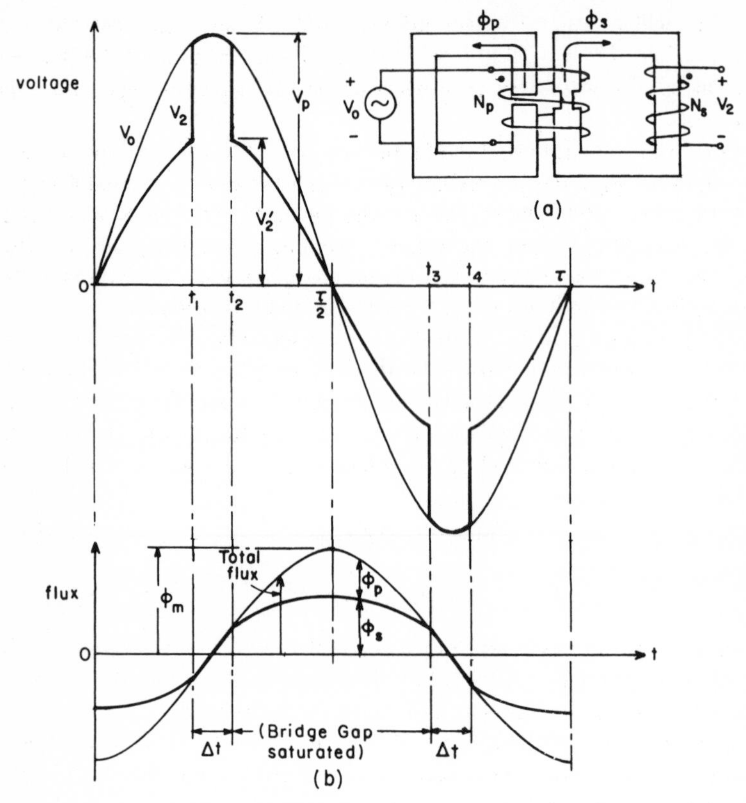

Figure 8.2 Structure for forming pulse on sine-wave voltage and waveforms.

8.2 Wroblewski Ballast

The Wroblewski ballast is an example of a regulating transformer that incorporates all of the elements described in the introduction to this chapter. The construction is shown in Figures 8.3 and 8.4. The transformer consists of a primary core and a series or secondary core, a primary winding over both cores, and a series or secondary winding. The primary core has a conventional air gap; the series core has one or two splined gaps.

The open-circuit output voltage consists of three components: a tap voltage on the primary winding, V_{0t}; an induced sine-wave series winding voltage, V_2'; and a pulse voltage, V_{11p}. Under load, the series winding also couples into the lamp circuit the inductive reactance produced by both the primary and series-core gaps.

The construction shown in Figure 8.4 utilizes a formed core, which is particularly suitable for spline-gap construction. The structure obviously can be

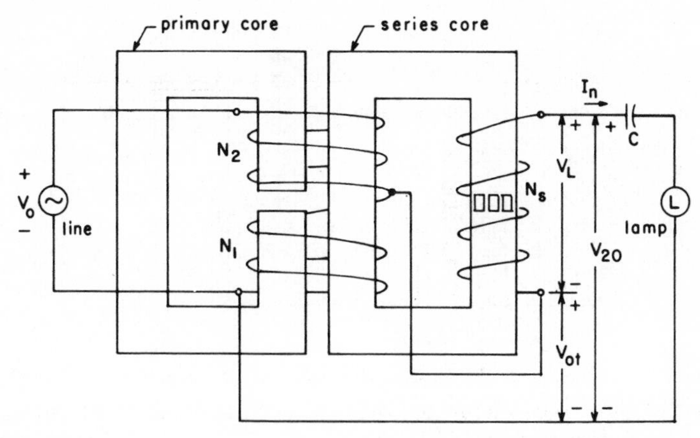

Figure 8.3 Magnetic circuit of Wroblewski regulating transformer.

built as an autotransformer with the tap voltage less or greater than the line voltage, and as an isolated winding transformer.

The fundamental-frequency equivalent circuit for the transformer is shown in Figure 8.5. The circuit consists of an ideal transformer having the physical

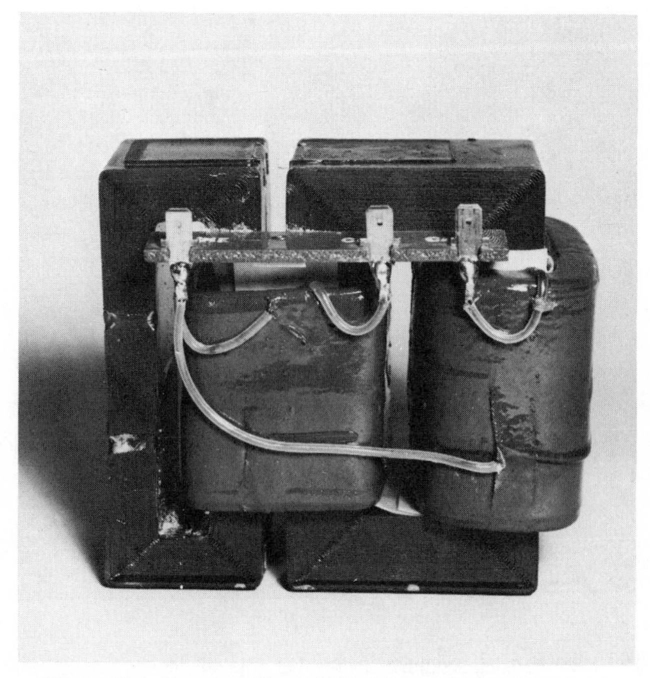

Figure 8.4 Construction of Wroblewski transformer.

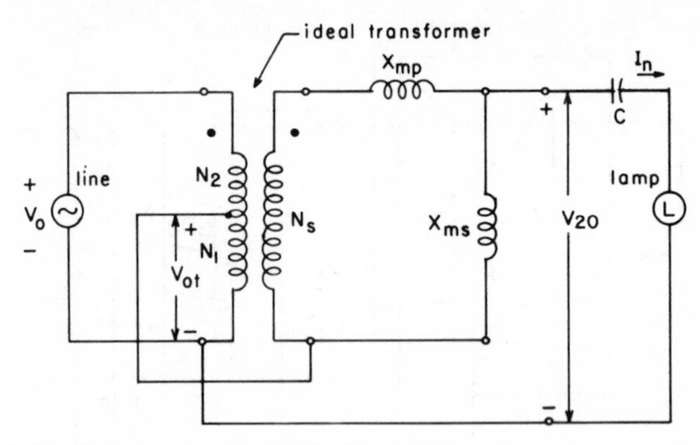

Figure 8.5 Equivalent circuit for Wroblewski structure as an autotransformer in ballast circuit.

turns ratio; a reactance X_{mp} corresponding to the magnetizing reactance of the primary core; and a reactance X_{ms} having the magnetizing reactance of the series core when the splines are saturated. The reduced equivalent circuit as seen by the lamp circuit is shown in Figure 8.6a. The source voltage, V_{0q},

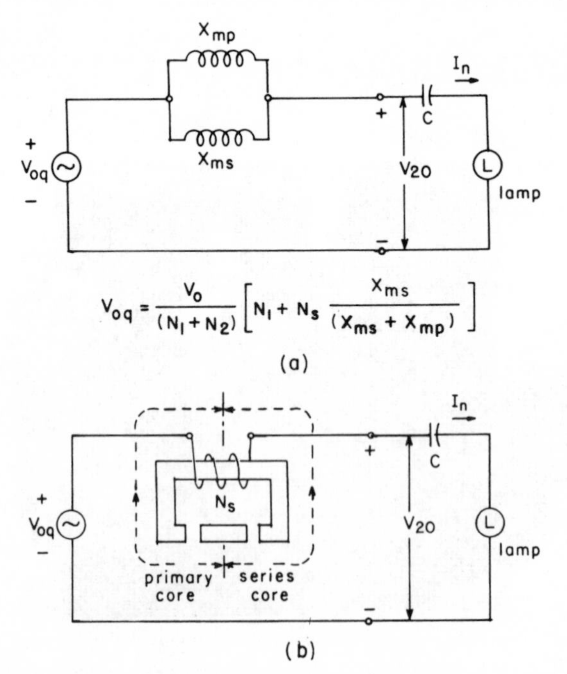

Figure 8.6 Reduced equivalent circuit for circuit of Figure 8.5: (a) circuit; (b) magnetic circuit analog.

is given by

$$V_{0q} = \frac{V_0}{(N_1 + N_2)} \left[N_1 + N_s \frac{X_{ms}}{(X_{ms} + X_{mp})} \right]. \tag{8.4}$$

The equivalent reactance

$$X_1 = \frac{X_{ms} X_{mp}}{(X_{ms} + X_{mp})}. \tag{8.5}$$

Reactances in an electric circuit are analogous to permeances in a magnetic circuit. Paralleled reactances are analogous to paralleled permeances or reluctances in series. Hence, the magnetic equivalent circuit of Figure 8.6b shows the two cores acting in series to produce the net reactance X_1. The reduced open-circuit equivalent circuit as seen by the supply line is shown in Figure 8.7a. The reactances are now in series so that the magnetic equivalent circuit of Figure 8.7b shows the cores acting on the N_s-turn winding in parallel.

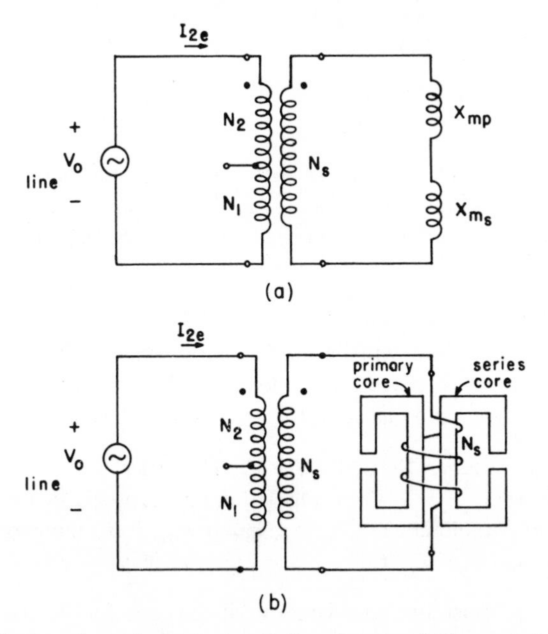

Figure 8.7 Reduced equivalent circuit for circuit of Figure 8.5 for determining magnetizing current: (a) circuit; (b) magnetic circuit analog.

The fundamental-frequency phasor diagram for the transformer is shown in Figure 8.8. The open-circuit fundamental-frequency voltage is given by the tap voltage V_{0t} plus the induced voltage V_2'. Under load, the series winding voltage because of the reactance becomes V_L and the total lamp circuit volt-

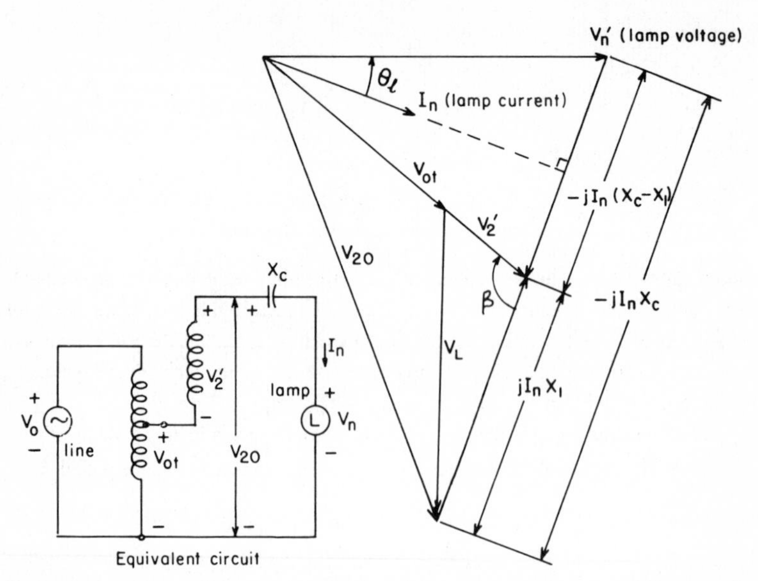

Figure 8.8 Phasor diagram for the transformer and circuit of Figure 8.5.

age increases to V_{20}. The diagram is used to find the required value of equivalent reactance X_1 which must be allocated to both cores, and the voltage V_L which determines the flux level at normal operation for the series core.

8.3 Design Procedure

The Wroblewski ballast will be used as an example of a regulating transformer which can be designed on a computer. The input variables to be selected by the designer and placed in the program are the following:

1. Lamp parameters: P_n, V_n, I_n, V_p, V_2', PFL_1.
2. Line parameters: V_0, f, regulation limits, power factor limits.
3. Transformer parameters: $H_0, k_i, k_\mu, V_{0t}, d, g_{0s}, J, B_s$, temperature limits.
4. Circuit parameters: C.

The computer must find the window dimensions for the two cores, the total width, the turns on each core, the wire sizes, the shunt-core gap, the losses, materials cost, temperature rise, and regulation.

The design variables of J, B_s, g_{0s}, can be reduced in number by substituting performance variables for which equations can be written or which can be handled by loops. The performance variables are line power factor, temperature rise and regulation. Using the performance variables, we would achieve either no design, or a unique design.

The design steps to be taken in the computer program are basically the following:

1. Waveform. Reduce the open-circuit waveform which consists of a pulse and a sine wave shown in Figure 8.2 to specific turns and core requirements of the magnetic circuit.
2. Electric Circuit. Solve the electric circuit for required reactances of the two cores, the voltages under load, and the currents.
3. Series Core. Design the series core to meet the reactance and voltage requirements previously determined.
4. Shunt Core. Design the shunt core and the primary winding to complete the magnetic circuit design.
5. Clean Up. Calculate the losses, temperature rise, material weight and cost, and regulation.

Rather than show an actual computer program a flow diagram is shown in Figure 8.9 and the key equations will be presented. The program defines the parameters of the two magnetic circuits; then proceeds to design them as two reactors using the procedure of Chapter 3.

A splined gap is employed in the series core to generate a voltage pulse. The design is based on the use of an equivalent butt gap for the splined gap to establish a constant (linear) reactance value. The permeance of the saturated splined gap can be considered as the sum of the permeance of the free gap plus the permeance of the saturated splines. The permeance of the free gap is given by

$$\mathcal{P}_{sg(\text{one gap})} = \mu_0 \frac{A_{gs}}{g_s'} (1 - k_i), \tag{8.6}$$

where

A_{gs} = area of the series core gap corrected for fringing,

g_s' = length of one series core gap corrected for core saturation,

k_i = per-unit interleave, e.g. 0.25 for one by four spline.

The permeance of the splines is given by

$$\mathcal{P}_{sb(\text{one gap})} = \mu_0 k_\mu \frac{A_{gs}}{g_s'} k_i, \tag{8.7}$$

where

k_μ = Relative permeability of the saturated spline. Typical values are 2-to-3 for k_i = 0.125 to 0.50.

The total reactance of the series core with n spline gaps in series is thus

$$X_{ms} = \frac{1}{n} (\mathcal{P}_{sg} + \mathcal{P}_{sb}) N_s^2 \, (2\pi f), \tag{8.8}$$

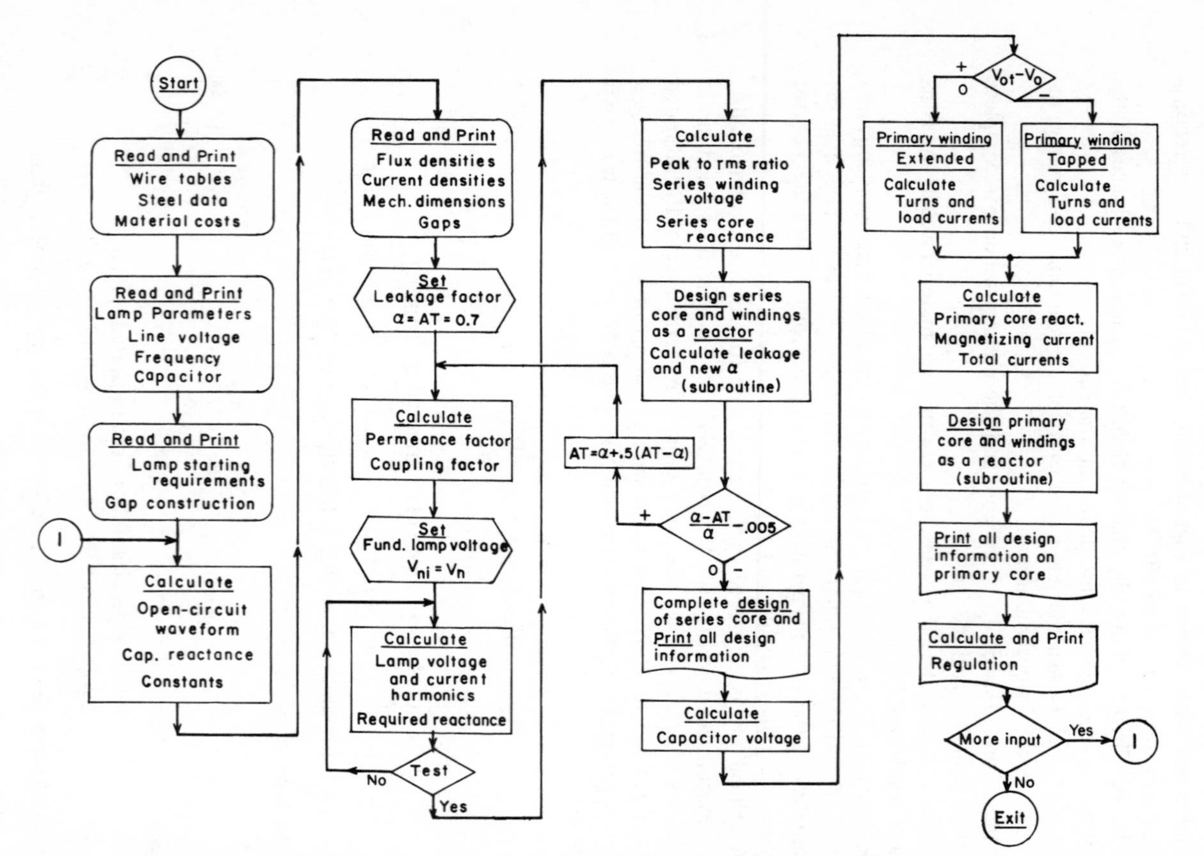

Figure 8.9 Flow diagram of computer program for regulating transformer design.

which becomes

$$X_{ms} = \frac{(2\pi f)}{n} N_s^2 \frac{A_{gs}}{g} \mu_0 \left[1 + k_i(k_\mu - 1)\right]. \tag{8.9}$$

The open-circuit voltage for the structure of Figure 8.3 is shown in Figure 8.10. The designer specifies the key parameters of the open-circuit voltage

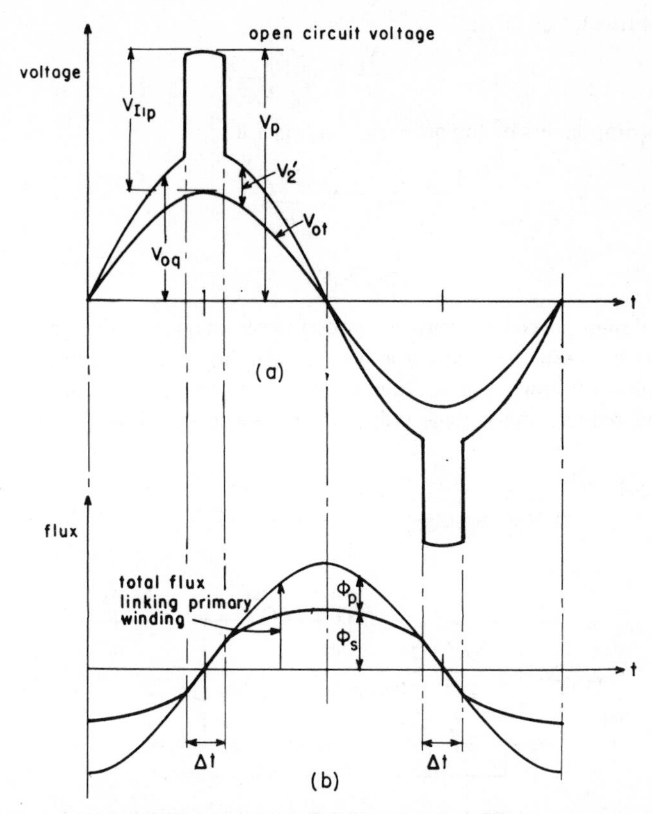

Figure 8.10 Waveforms for Wroblewski transformer: (a) open-circuit output voltage; (b) flux.

from his knowledge of the lamp requirements and the program proceeds to design the ballast that will provide the voltage. The required parameters to define the design are

1. V_p = peak open-circuit voltage.
2. V_{0t} = rms tap voltage.
3. k_i = per-unit spline of series core.
4. V_2' = sine-wave part of induced voltage, rms.
5. V_0 = line voltage.

These input parameters will set the primary-to-secondary turns ratio, N_0 to N_s, and the permeance coefficient, k_p, which is the ratio of primary core permeance to series core permeance, $\mathcal{P}_s/\mathcal{P}_p$.

The induced pulse voltage is given by

$$V_{11p} = V_p - \sqrt{2}\, V_{0t} \,. \tag{8.10}$$

The turns ratio is then

$$\frac{N_s}{N_0} = \frac{V_{11p}}{\sqrt{2}\, V_0} \,. \tag{8.11}$$

The components of the primary winding are

$$N_1 = N_0 \frac{V_{0t}}{V_0} \,, \tag{8.12}$$

$$N_2 = N_0 - N_1 \,. \tag{8.13}$$

The design procedure must account for the magnetic leakage or imperfect coupling between the primary and series winding because of flux crossing the series core window. The leakage adds to the reactance produced by the air gaps and reduces the voltage induced into the series winding by the primary flux.

The equivalent circuit of Figure 8.5 is modified to the form shown in Figure 8.11. A leakage reactance X_{sl} is placed in series with the reactance

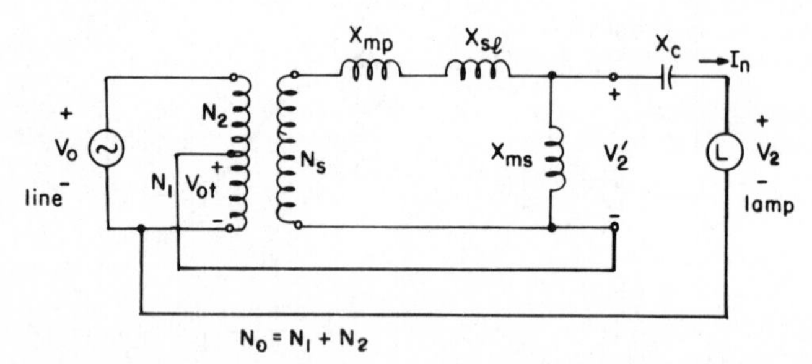

Figure 8.11 Equivalent circuit showing leakage reactance between primary and series winding.

X_{mp} of the primary core. Two coefficients describing the magnetic structure are now defined

$$\text{permeance ratio, } k_p = \frac{X_{ms}}{X_{mp}} = \frac{\mathcal{P}_s}{\mathcal{P}_p} \,, \tag{8.14}$$

$$\text{leakage factor, } (1 - \alpha) = \frac{X_{sl}}{X_{ms}} = \frac{\mathcal{P}_l}{\mathcal{P}_s} \,. \tag{8.15}$$

The sine-wave part of the induced voltage is

$$V_2' = V_0 \frac{N_s}{N_0} \frac{X_{ms}}{X_{ms} + X_{sl} + X_{mp}}, \tag{8.16}$$

which reduces to

$$V_2' = V_0 \frac{N_s}{N_0} \frac{1}{1/k_p - \alpha + 2}. \tag{8.17}$$

Equation 8.17 can be solved for k_p.

$$k_p = \frac{1}{(V_0/V_2')(N_s/N_0) + \alpha - 2}. \tag{8.18}$$

Using Equation 8.11, we find that

$$k_p = \frac{1}{(V_{11p}/\sqrt{2}\, V_2') + \alpha - 2}. \tag{8.19}$$

Equation 8.19 for the permeance ratio k_p determines the geometry and gap allocation between the two cores in terms of the given input voltage components V_{11p} and V_2'. Just as in the procedure for the reactor in Chapter 3, the leakage factor α is assumed at the start of the design sequence and then calculated from the dimensions of the completed design. The design sequence is repeated until the value of α stabilizes. Typical value for α is 0.7 for the two-core structure.

The permeance coefficient k_p also enters into the expression for the net reactance X_1 seen by the lamp and the capacitor in the circuit of Figure 8.11. The net reactance is

$$X_1 = \frac{X_{ms}(X_{mp} + X_{sl})}{X_{ms} + X_{mp} + X_{sl}} \tag{8.20}$$

$$= \frac{X_{ms}(1 + k_p - k_p\alpha)}{(1 + 2k_p - k_p\alpha)}. \tag{8.21}$$

The fundamental-frequency open-circuit voltage is given by

$$V_{0q} = V_{0t} + V_2', \tag{8.22}$$

which can be written in terms of a coupling coefficient k_c as

$$V_{0q} = V_{0t} + V_0 \frac{N_s}{N_0} \frac{k_p}{(k_p + 1)} k_c, \tag{8.23}$$

where

$$k_c = \frac{k_p + 1}{(2k_p - k_p\alpha + 1)}. \tag{8.24}$$

For no leakage, $\alpha = 1$ and the coupling coefficient $k_c = 1$.

The phasor diagram for the fundamental-frequency operation of the transformer is shown in Figure 8.8. Using the diagram, the given values of the lamp parameters, and the open-circuit voltage V_{0q} from Equation 8.22, the reactance voltage $I_n (X_c - X_1)$ can be found, and from it the required ballasting reactance X_1, using the methods of Chapter 7.

The fundamental voltage appearing across the series winding N_s at no load is V_2', but increases to V_L under load. The value can be found from the phasor diagram as

$$V_L = [(V_2')^2 + (I_n X_1)^2 + 2V_2'(I_n X_1) \cos \beta]^{1/2}, \tag{8.25}$$

where

$$\cos \beta = \left[1 - \frac{(V_{n1}')^2 \, (\mathrm{PFL}_1)^2}{V_{0q}{}^2}\right]^{1/2} \tag{8.26}$$

As in the design of a reactor in Chapter 3, the worst case voltage will include the third-harmonic voltage V_{n3} of the lamp. The worst case voltage V_L' is thus

$$V_L' = V_L + 0.33\, V_{n3} \frac{3X_1}{(3X_1 - X_c/3)}. \tag{8.27}$$

The series-core magnetic circuit can be designed using the computer procedure of Chapter 3 for a reactor. The required electrical parameters for the series core are obtained from the following equations:

1. Reactance, X_{ms}, from Equation 8.21.
2. Current, I_n, from lamp specification.
3. Maximum voltage, V_L', from Equation 8.27.

At the conclusion of the series-core design, the permeance of the leakage paths is calculated, and from it the leakage reactance X_{sl}. A new value of leakage coefficient α is calculated and the design calculations repeated if necessary as shown on the flow diagram. The splines are handled by using Equation 8.9 for the reactance of an equivalent butt-gap core. The per-unit interleave k_i is specified by the designer to obtain a prescribed pulse width. Factor k_μ is a second-order correction and is estimated.

The turns of the primary winding are given by Equations 8.11, 8.12, and 8.13 from the input variables and the already determined series core winding N_s. The core geometry and gap of the primary core are calculated from Equation 8.20. The equation is solved for the primary core reactance X_{mp} referred to the N_s-turn winding, and the primary core permeance $\mathcal{P}_p$ calculated.

The remaining variables to be determined for the primary core are the winding currents to determine the wire sizes and window size. The total current I_{1t} and I_{2t} in the two primary winding sections can each be considered as having two components

$$I_{1t} = I_1 + I_{m1}, \tag{8.28}$$

$$I_{2t} = I_2 + I_{m2}, \tag{8.29}$$

where I_1 and I_2 are the components to produce ampere-turn balance with the lamp current I_n, and I_{m1} and I_{m2} are the primary magnetizing currents. The summation is vectorial.

The component I_2 is given by

$$I_2 = I_n \frac{N_1}{N_0} + I_m \frac{N_s}{N_0} \frac{k_p}{(2k_p - \alpha k_p + 1)}, \tag{8.30}$$

and

$$I_1 = I_2 - I_n. \tag{8.31}$$

The magnetizing current is given by

$$I_{m1} = I_{m2} = \frac{V_0}{\frac{N_0^2}{N_s^2} X_{ms} \left(\frac{2k_p - k_p \alpha + 1}{k_p} \right)} = I_m. \tag{8.32}$$

The phasor diagram showing the currents is given in Figure 8.12. From the diagram, we see that the total currents are

$$I_{2t} = [I_2^2 + I_m^2 - 2I_2 I_m \cos \alpha]^{1/2}, \tag{8.33}$$

$$I_{1t} = [I_1^2 + I_m^2 - 2I_1 I_m \cos \alpha]^{1/2}, \tag{8.34}$$

where

$$\cos \alpha = \left[1 - \frac{V'_{n1} \, \mathrm{PFL}_1^2}{V_{0q}} \right]^{1/2}. \tag{8.35}$$

With the currents and winding turns, the ampere-turns and winding space can be calculated and the design completed.

The regulation of lamp power P_n versus line voltage V_0 is calculated by an iterative process from the phasor diagram of Figure 8.8. The circuit regulates by limiting the increase of lamp current I_n through a decrease of reactance X_1, and an increase of reactance $(X_c - X_1)$, when the line voltage is raised. The reactance X_1 corresponds to the voltage V_L and V'_L which appears across the series winding. The voltage only affects the X_{ms} part of the reactance. Hence, the procedure finds for each step of line voltage V_0 from nominal, a new combination of lamp current I_n, series-core flux density, voltage V_L and reactance X_1, which satisfies the phasor diagram. The lamp voltage V_n and PFL_1 are assumed to remain fixed. The lamp power for each new line voltage point is then calculated as

$$P_n = V_n I_n \mathrm{PFL}_1. \tag{8.36}$$

Table 8.1 Typical Printout of the Design of a 250-W Wroblewski Type Regulating Ballast Transformer

```
C-102 250 WATT 240 V 60 CS HG 1/30/67 HIGH AMBIENT
CASE NO     4
DESIGN OF SERIES CHOKE
    BW        CJ         GOS       VMR    IRON COST
128.0000  1800.0002    0.0625   0.3000     0.1820
HO = 3.3750       DLIM = 1.75000
ANL =  691.31        AN2 =   691.32

BUILD = 0.68969
WIRE NO                 18
TOTAL TURNS            642
TURNS PER LAYER         45
NUMBER OF LAYERS  14.25

IRON WT = 2.26 LBS    COPPER WT = 1.89 LBS   COPPER RESISTANCE = 2.41 OHMS
CORE LOSS = 8.87 WATTS  COPPER LOSS = 12.77 WATTS  TOTAL LOSSES = 21.65 WATTS
IRON COST = $0.411         COPPER COST = $1.391       TOTAL COST = $ 1.80

             SERIES CORE AND COIL DIMENSIONS
     HO          WO          XL          H1          W1          D          DP
   3.3750     2.3741      0.5312      2.3125      1.3116      1.7500      3.1909

NC =  5    T1 = 72.76         T1C = 72.76
A1 = 0.835    A2 = 0.837    CAPACITOR VOLTS = 254.72    KP = 0.50311E 00

DESIGN OF PRIMARY CHOKE
   BM2        CJ2        HO2        WS        VMR        IRON COST
 100.00  1200.00      3.3750    0.3125    0.3000       0.1820
N = 2    A2 OLD = 0.974E 00    A2 NEW = 0.974E 00

PRIMARY GAP =   0.0205
COST OF COPPER, 1ST WINDING, = $ 0.784 COST OF COPPER, 2ND WINDING, = $ 0.642

BUILD FOR N1 COIL = 0.5977       BUILD FOR N2 COIL = 0.6191
                        N1                  N2
WIRE NO                 18                  19
TOTAL TURNS            326                 326
TURNS PER LAYER         25                  22
NUMBER OF LAYERS   13.00               14.75
COPPER WT             1.07 LBS            0.85 LBS
COPPER LOSS           3.23 WATTS          2.59 WATTS
RESISTANCE            1.36 OHMS           1.72 OHMS

CORE LOSS    2.77 WATTS    TOTAL LOSS    8.60 WATTS
IRON COST = $ 0.249        COPPER COST = $ 1.427    TOTAL COST = $ 1.67

                     PRIMARY CORE AND COIL DIMENSIONS
     HO          WO          XL          H1          W1          D          DP
   3.375       1.423       0.371       2.631       0.680      1.750       3.125

    XM          IM          I1         I2        I1T        I2T        XMS        X1          X2
202.5423   1.1849   -0.4312   1.8687   1.5406   1.2249   62.3402   42.5607   127.7296
```

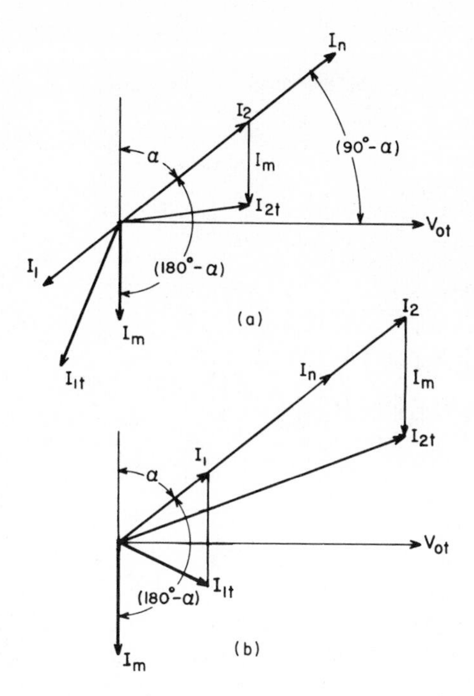

Figure 8.12 Phasor diagram for primary currents: (a) $I_n > I_2$; (b) $I_n < I_2$.

8.4 Printout Form

A printout form for the computer design of a 250-W 240-V mercury-vapor lamp ballast is shown in Table 8.1. The computer program uses subroutines originally written for reactors to design the series and primary cores. The printout shows the design of each of the cores separately. The printout also gives the electrical information accumulated during the solution of the electric circuit, such as the reactance components, current components and voltages.

8.5 Summary

The chapter points out that two requirements must be fulfilled to obtain a successful computer design procedure for a complex magnetic circuit. First, the physical magnetic circuit must be modeled in sufficient detail to insure that the dimensions and turns found on the computer will yield the reactance values desired. Second, the electric circuit equations must be derived so that the computer can solve them for all of the currents, voltages, and reactances that define the magnetic circuit. Once these two requirements are fulfilled the design of the magnetic circuits themselves is straightforward.

9 MULTIWINDING TRANSFORMERS

The transformers in this chapter are "ordinary" power-frequency types having straightforward magnetic circuits. The purpose of the computer program is to save design time and keep the parameters organized. The output forms can be set up so that the output data can be used for preparing quotations and for preparing manufacturing instructions.

9.1 Multiwinding Transformers

The magnetic circuit to be used is a transformer having one or more primary windings and one or more secondary windings. The operating frequency is in the power range 50 Hz, 60 Hz, and 400 Hz, and the size may extend up to a few kVA. Typical applications are control-circuit transformers, electronic-equipment power supplies, and lighting transformers.

The secondary loads are assumed to be resistive and the winding regulation calculated arithmetically from the winding resistance. The leakage reactance is neglected as a factor in the regulation. The flux density is selected to meet loss and temperature rise requirements. The magnetizing current is neglected.

The magnetic-circuit design portion of the program has only two objectives. The first is to make the windings fit into the window; the second is to make the temperature rise fall into a prescribed narrow range. The input variables bind the design freedom of the computer.

The remainder of the program has the objective of working out the winding and insulation arrangement so that the space requirement is accurately calculated and the amounts of paper, wire, and core material are accurately calculated. The printout is very detailed with cost and manufacturing information.

9.2 *Inputs*

The input information for the program is the following:

1. Project number and transformer type.
2. Wire table, twenty consecutive sizes suitable for the design, including size, weight, resistance, required interlayer paper, and cost.
3. Steel table, shows losses in watts per pound from 60,000 to 120,000 lines per square inch in 10,000 intervals, and price per pound.
4. Paper, core tube and tape sizes and price.
5. Core information, strip width, length and width of core section, type of core, surface area of case if case is used, initial value of flux density.
6. Number of secondary windings, winding order, primary voltage, frequency and maximum temperature rise.
7. Secondary voltage, current, form factor for rectifier loads, regulation and tolerance.
8. Secondary current density, area factor of core, margin, primary current density.

The computer must select the window dimensions, the wire sizes, the number of turns and the layout of insulating paper and tape.

9.3 *Design Procedure*

The steps of the computer program are shown in the flow chart of Figure 9.1. The key steps are explained as follows:

1. Voltamperes

$$\text{Rated VA} = \text{VA} + \Sigma \text{ES} \times \text{AMP} \times \text{FAC}, \tag{9.1}$$

where

VA is initially set to zero,

ES = secondary voltage, each winding,

AMP = secondary current,

FAC = DC to RMS conversion factor for rectifier loads.

2. Primary amperes

$$\text{PRA} = 1.10 \times \frac{\text{VA}}{\text{EP}}, \tag{9.2}$$

where

EP = rated primary voltage,

1.10 = 10% assumed losses.

3. Core "centerleg" area

$$\text{AREA} = \text{FFAC} \times \frac{60}{\text{FREQ}} \times \sqrt{\text{VA}}, \tag{9.3}$$

where

FFAC = area factor (to establish steel to copper ratio).

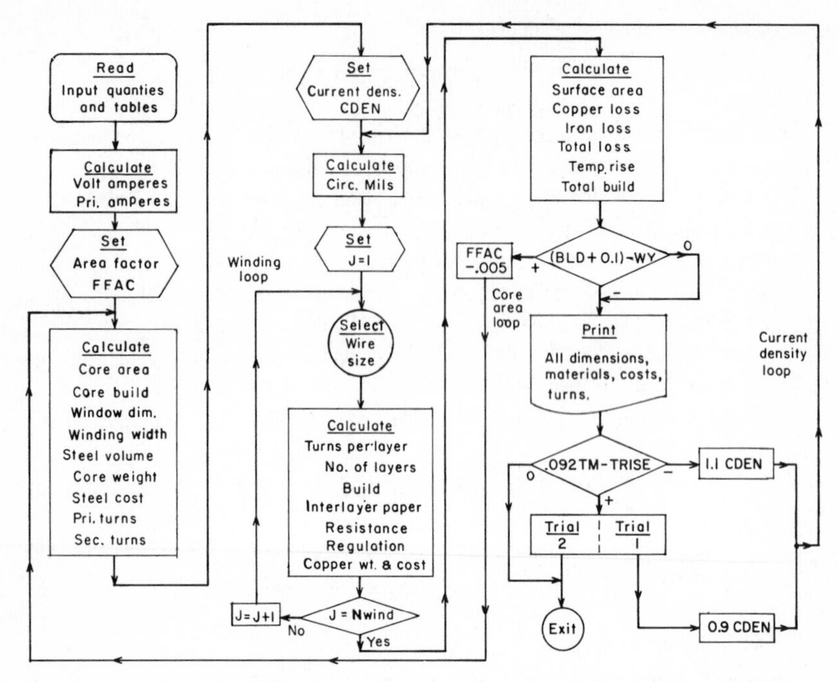

Figure 9.1 Flow diagram of computer program for multiwinding transformer design.

4. Core build

The program is based upon a formed core made from strip material. The steps are shown for the shell-type core made from two core units.

$$\text{CB} = \text{AREA} / 2 \times \text{CSTW}, \tag{9.4}$$

where

$$\text{CSTW} = \text{core strip width}.$$

5. Core centerleg perimeter

$$\text{CCF} = 2 \ \ [2 \times \text{CB} + \text{CSTW})]. \tag{9.5}$$

6. Window

See Figure 9.2 for terminology.

$$\text{WX} = \text{CORL} - (2 \times \text{CB}), \tag{9.6}$$

$$\text{WY} = \text{CORW} - (2 \times \text{CB}). \tag{9.7}$$

7. Winding width

$$\text{WTH} = \text{WX} - \text{EDGE}, \tag{9.8}$$

where

$$\text{EDGE} = \text{total margin} + \text{core tube clearance}.$$

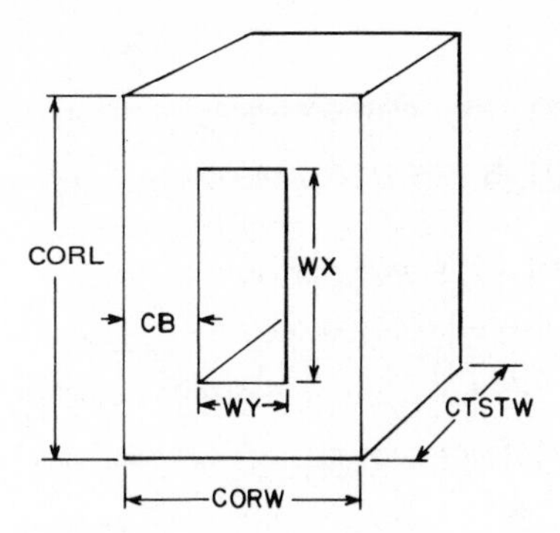

Figure 9.2 One transformer core.

8. Volume of steel

$$FEV = CSTW\ [CORL \times CORW - (WX \times WY)] \times 2. \tag{9.9}$$

9. Core weight

$$CLB = FEV \times 0.26, \tag{9.10}$$

where

0.26 = density at 0.93 stacking factor.

10. Steel cost

$$FECST = 1.00 \times CLB \times PRION \times 1000, \tag{9.11}$$

where

PRION = price per pound,

1.00 = 100 per cent usage,

1000 = price for 1000 sets.

11. Primary turns

$$NTRN\ (1) = \frac{EP \times 10^8}{4.1293 \times AREA \times FREQ \times BMAX}, \tag{9.12}$$

where

4.1293 = 4.44 × 0.93 (stacking factor),

BMAX = lines per square inch.

12. Secondary turns

Each secondary taken in order of winding.

$$NTRN\ (L) = \frac{NTRN\ (1) \times ES \times (1. + REG)}{EP}, \tag{9.13}$$

where

REG = regulation defined as input.

13. Wire size

$$\text{CMIL (1)} = \text{PDEN} \times \text{PRA}, \tag{9.14}$$

$$\text{CMIL (L)} = \text{CDEN} \times \text{AMP (L)}, \tag{9.15}$$

where

CMIL (1) = wire size of primary,

PDEN = current density of primary,

PRA = computed primary current,

CMIL (L) = wire size of secondaries,

CDEN = current density of secondaries,

AMP (L) = current in respective secondary.

14. Turns per layer

$$\text{NTPL (L)} = \frac{0.95 \times \text{WTH}}{\text{DIAM (I)}}, \tag{9.16}$$

where

0.95 = winding space factor,

WTH = winding width,

DIAM = respective wire diameter.

15. Number of layers

$$\text{NLAY (L)} = \frac{\text{NTRN (L)}}{\text{NTPL (1)} + 1}. \tag{9.17}$$

Extra turn to provide for lead length for coil finish.

16. Coil wrap (interwinding)

$$\text{WPAP} = \text{WPAP} + 3 \times (6.5 \times \text{BLD} \times \text{WX} + \text{CCF}), \tag{9.18}$$

where

WPAP = wrap paper (= zero at start),

3 = no. of layers of 0.005-in Kraft for each winding wrap,

6.5 = $2\pi \times 1.035$ (correction for winding machine),

BLD = buildup due to previous winding.

17. Build

$$\text{BLD} = 0.050 + (\text{XN}) \times \text{DANDP (I)} - 0.015, \tag{9.19}$$

where

0.050 = build due to core tube and clearance,

XN = no. of layers − 1,

DANDP = diameter of wire and paper,

- 0.015 = 3 layers of 0.005-in paper which are not used over core tube.

18. Paper size (interlayer)

$$\text{PSIZE (L)} = \text{DANDP (I)} - \text{DIAM (I)}. \qquad (9.20)$$

Paper is listed with respective wire in table. Subtracting the wire diameter from DANDP leaves the paper thickness.

19. Resistance per winding

$$\text{RES (L)} = \text{ALT (L)} \times \text{TRN} \times \text{OHMK (I)} / 12000, \qquad (9.21)$$

where

ALT = average length of turn-inches,

OHMK = ohms per 1000 feet,

TRN = number of turns.

20. Resistance at operating temperature

$$\text{RET (L)} = \text{RES (L)} \ [1 + 0.00393 \times (\text{TMAX} + 20)], \qquad (9.22)$$

where

0.00393 = temperature coefficient of copper,

TMAX = allowable temperature rise.

21. Voltage drop at operating temperature

$$\text{VOLT} = \text{RET (L)} \times \text{AMP (L)}. \qquad (9.23)$$

22. Regulation at operating temperature

$$\text{FREG (L)} = \text{VOLT}/[\text{ES (L)} \times (1.0 + \text{REG (L)} - \text{VOLT})]. \qquad (9.24)$$

23. Interlayer paper

$$\text{SQINP (L)} = \text{WX} \times \text{ALT (L)} \times \text{XN}. \qquad (9.25)$$

24. Copper weight

$$\text{WNDWT} = \text{ALT (L)} \times \text{TRN} \times \text{WTK (I)} / 12000, \qquad (9.26)$$

where

WTK (I) = wire weight per 1000 feet.

25. Copper cost

$$\text{CUCST} = \text{WNDWT} \times 10.0 \times \text{PRCE}, \qquad (9.27)$$

where

PRCE = price of wire per 100 pounds.

10.0 = factor to convert to price per 1000 units.

26. Surface area
For open core and coil.

$$\begin{aligned} \text{SURF} = 2 \times \text{CSTW (CORL} + 2\ \text{CORW)} + \\ 2\ (\text{CORL} \times 2\ \text{CORW}) + \\ 3\ \text{WY}\ \ [(2\text{CB} + 1.5\ \text{WY}) + \text{WX}]. \end{aligned} \tag{9.28}$$

27. Copper loss

$$\text{WATT (L)} = \text{RET (L)} \times \text{AMP (L)}^2 \tag{9.29}$$

28. Iron loss
This formula interpolates for any value between 60 and 120 kilolines per square inch.

$$\text{WPP} = \text{T (M)} + [\text{T(M} + 1) - \text{T(M)} \times (\text{X} - \text{Y})], \tag{9.30}$$

where

WPP = watts per pound,

T = flux density in 10 kiloline per square inch intervals,

$\text{X} = \text{BMAX}/10{,}000$,

$\text{M} = \text{X} - 5$,

$\text{Y} = \text{X} + 5$.

29. Total iron loss

$$\text{FEW} = \text{CLB} \times \text{WPP}. \tag{9.31}$$

30. Total losses

$$\text{TLOSS} = \text{TWATT} + \text{FEW}. \tag{9.32}$$

31. Temperature rise

$$\text{TRISE} = \text{TLOSS} \times .65/\text{SURF}. \tag{9.33}$$

32. Temperature rise adjustment

This computation adjusts the input current density to meet but not exceed a temperature rise of 92 per cent of maximum.

If

$$\begin{aligned} &0.92 \times (\text{TMAX}) - \text{TRISE} = 0, \text{ end sequence}, \\ &0.92 \times (\text{TMAX}) - \text{TRISE} > 0, \text{ use } 0.9 \times \text{CDENS}, \\ &0.92 \times (\text{TMAX}) - \text{TRISE} < 0, \text{ use } 1.1 \times \text{CDENS}. \end{aligned} \tag{9.34}$$

33. Coil build adjustment
This computation checks the coil build with the window space. If the build is too large, the window area is increased.

Table 9.1 80-Column Card Layout for Input Data for Multiwinding Transformer Program. Bracketed Items Are Descriptive and Are Not Punched. Total Deck of 27 Cards

CARD NO.									
		(Project Description)	(Varnish)	(Kraft paper)	(Tape)	(Steel Thickness)	(Steel Grade)	(Date)	(Magnet Wire)
1	1	PROJECT 30542 MACH TOOL-4	MB30	K	MM3	.0185	M-22	12/27/67	SS
	(Wire size)	(Area Circular Mils)	(Ohms/1000 ft)	(Weight/1000 ft)	(Wire Diameter)	(Wire Dia.+ paper)	(Wire Price)	(Paper Price)	
2	11	8227.00	1.2610	25.1000	.0927	.1027	50.55	.1400	
TO	TO								
21	30	100.00	103.71	.3110	.0110	.0130	74.15	.0375	
	(Loss - Watts per Pound) (60Kℓ)	(70Kℓ)	(80Kℓ)	(90Kℓ)	(100Kℓ)	(110Kℓ)	(120Kℓ)	(Steel Price)	
22	.7	.9	1.2	1.5	1.8	2.1	2.7	.1820	
	(Paper Price)	(Price Core Tube)	(Price 3/4-in Tape)	(Price 1/2-in Tape)					
23	.0935	.007	.675	.465					
	(Core Strip Width)	(Core Length)	(Core Width)	(Surface Area)	(Induction Lines/sq.inch)				
24	1.750	2.875	1.843	0.000	102000.				
	(Number of Pri Sec)	(Pri Volts)	(Frequency)	(Max. Temp.Rise)					
25	1	2440.	60.0	60.0					
	(Sec. Volts)	(Sec Load Current)	(Factor)	(Regulation)	(Tolerance)				
26	115.0	1.300	1.030	.060	.030				
	(Sec. Current Density)	(Core Area Factor)	(Margin)	(Pri Current Density)	(Check Regulation)				
27	620.	.158	.3125	430.	.200				

If

$$\begin{aligned} &\text{CKBLD} + 0.100 = \text{WY, end sequence,} \\ &\text{CKBLD} + 0.100 < \text{WY, end sequence,} \\ &\text{CKBLD} + 0.100 > \text{WY, use FFAC} - 0.005. \end{aligned} \tag{9.35}$$

34. Determination of tape width

This computation selects 3/4-in tape for wire size 16 or larger diameter and 1/2-in tape for wire size 17 or smaller diameter.

Table 9.2 Typical Printout Form for a 150-VA Control Transformer Having Two 220-V Primary and One 115-V Secondary Windings

PROJECT 30542 MACH TOOL-4 M830 K MM3 .0185 M-22 12/27/67 SS

WINDING ORDER	2	3	1
WIRE SIZE	28	25	21
NO. OF TURNS	417	417	231
TURNS PER LAYER	96	68	43
NO. OF LAYERS	5	7	6
PAPER SIZE	0.0020	0.0029	0.0030
WINDING WIDTH	1.375		
ARBOR SIZE			
RESISTANCE	17.83	9.79	1.68
WEIGHT OF WIRE	0.134	0.297	0.327
WATTS CU	3.26	7.17	3.52
FINAL REGULATION	0.040	0.044	0.022
TOTAL WATTS CU	13.96		
SECONDARY VA	153.98		
AREA FACTOR	0.168		
WINDOW LENGTH	1.68		
WINDOW WIDTH	0.65		
TOTAL BUILD	0.525		
SURFACE AREA	51.7		
WATTS STEEL	7.09		
TEMPERATURE RISE	67.1		
WEIGHT OF CORE	3.81		
STEEL COST/1000 $	694.35		
WIRE COST/1000 $	93.30	192.35	191.25
SQ. IN. PAPER	66.27	102.86	69.45
PAPER COST/1000 $	2.48	5.91	3.99
WRAP PAPER AREA	90.54		
WRAP COST/1000 $	8.46		
CORE TUBE/1000 $	11.81		
LGTH 3/4 IN TAPE	0.00		
3/4 IN COST/1000	0.00		
LGTH 1/2 IN TAPE	10.12		
1/2 IN COST/1000	4.70		
TOTAL COST/1000 $	1208.64		

If

$$\text{wire size} - 16 < 0, \text{ use } \tfrac{3}{4}\text{-in tape,}$$
$$\text{wire size} - 16 = 0, \text{ use } \tfrac{3}{4}\text{-in tape,} \qquad (9.36)$$
$$\text{wire size} - 16 > 0, \text{ use } \tfrac{1}{2}\text{-in tape.}$$

35. Tape length

$$\text{TPBG} = \text{TPBIG} + 2 \times \text{WX}. \qquad (9.37)$$

36. Tape cost

$$\text{COST} = \text{Sum of all holding tapes} \times \text{PRICE}. \qquad (9.38)$$

37. Core tube cost (1000 pieces)

$$\text{CTCST} = 1000 \times \text{WX} \times \text{CTPRC}, \qquad (9.39)$$

where

$$\text{CTPRC} = \text{cost per inch.}$$

9.4 Input-Output Forms

The input to a design program is usually prepared on cards. The output is presented as a printout of all of the pertinent information.

A sample of an input card for the design of a 150-VA machine-tool control transformer is shown in Table 9.1. The various items are labeled and are explained in Section 9.3. Card 2 is one of the 20 prepunched cards for 20 wire sizes and is not prepared for each design.

Figure 9.3 Typical transformer designed by program of Chapter 9.

A sample of the printout for the input card is shown in Table 9.2. The items are self-explanatory. The amount and cost of the materials is prepared completely, as is the performance. The transformer can be manufactured from the combined input and output data. A typical transformer designed with this program is shown in Figure 9.3.

The program is very constrained and has very little freedom for optimization. The program prepares a design with full information to fit the outer core size prescribed.

TABLE OF SYMBOLS

For electrical quantities, lower case symbols are used for instantaneous values; upper case symbols are used for rms, average or peak values.

A_g = cross-sectional area, air gap, corrected for fringing
A_m = cross-sectional area, core, effective
A_w = cross-sectional area, window

B_g = flux density, air gap
B_m = flux density, core
B_s = flux density, saturation

C = capacitance

d = depth, core

f = frequency, line
g_0 = length, mechanical air gap
Δ_g = gap length correction for saturation
g' = length, equivalent air gap

H_m = magnetic field intensity
H_0 = height, core
H_1 = height, window, per winding

I_m = magnetizing current
I_{m1}, I_{m2} = magnetizing current for core sections 1, 2
I_{mp}, I_{ms} = magnetizing current for primary and series cores
I_n = current, load or lamp
I_{nk} = kth harmonic of current ($k = 1, 3, 5, \ldots$), rms value
I_{p1}, I_{p2} = current, primary windings N_1, N_2
I_0 = current, line
I_1, I_2 = current, primary and secondary windings

I_{1t}, I_{2t} = current, total, primary winding sections of Wroblewski transformer
J = current density
k_c = coupling coefficient
k_i = per-unit interleave of spline gap
k_p = permeance ratio
k_q = ratio of capacitive to series reactance
$k\mu$ = relative permeability
L = inductance
L_g = inductance component for air gap
L_m = inductance component for core
L_s = inductance component from non-gap leakage fields
l_m = magnetic path length, core
M_r = coil margin
N = turns
N_1, N_{p1} = turns, primary winding, below tap
N_2, N_{p2} = turns, primary winding, above tap
N_s = turns, series, secondary winding
Ni_g = ampere-turns, gap
Ni_m = ampere-turns, core
P_n = lamp power
P_0 = line power
P_p = power rating, primary
P_q = reactive power
P_s = power rating, secondary
P_t = power rating, total
$\mathcal{P}$ = permeance
$\mathcal{P}_g$ = permeance, air gap
$\mathcal{P}_l$ = permeance, leakage path
$\mathcal{P}_m$ = permeance, core
$\mathcal{P}_p$ = permeance, primary core
$\mathcal{P}_s$ = permeance, series core
$\mathcal{R}$ = reluctance
T = temperature
t = time
V_L = voltage, reactor or series winding, under load
V_{Lk} = voltage, kth harmonic ($k = 1, 3, 5, \ldots$), rms value
V_l = voltage, leakage reactance
V_{I1p} = pulse voltage amplitude, open circuit
V_n = voltage, lamp or load

V'_n = voltage, corrected for circuit loss
V_{nk} = voltage, kth harmonic ($k = 1, 3, 5, \ldots$), rms value
V_0 = voltage, line
V'_0 = voltage, equivalent source
V_{0h}, V_{0l} = line voltage, upper and lower limits
V_{0q} = voltage, total open-circuit output
V_{0t} = tap voltage, primary winding
V_p = peak output voltage, open circuit
V'_1 = equivalent voltage, reactor
V'_2 = voltage, series winding, open circuit
V_{20} = voltage, total output, under load
$\mathcal{V}_m$ = volume of core
$\mathcal{V}_g$ = volume of air gap

w = energy density
W = energy, total
W_0 = width, core
W_p = energy, peak
W_1 = width, window

X_c = reactance, capacitor
X_L = reactance, series
X_l = reactance, leakage
X_{ll}, X_{lh} = reactance, low and high line-voltage
X_m = reactance, magnetizing
X_{m1}, X_{m2} = reactance, magnetizing, core sections 1, 2
X_{mp}, X_{ms} = reactance, magnetizing, primary, series cores
X_1 = reactance, total series
X_1 = width, core leg

α = leakage coefficient
θ_l = power factor angle, lamp, $\cos \theta_l = \mathrm{PFL}_1$
θ_0 = power factor angle, line
μ_0 = permeability, free space
μ_m = permeability, core
ϕ = magnetic flux
ϕ_e = leakage flux, external
ϕ_l = leakage flux, between windings
ϕ_p, ϕ_s = flux, primary, series, cores
ϕ_{w1} = leakage flux, across windows through windings
ϕ_{w2} = leakage flux, across windows, between windings
ϕ_1, ϕ_2 = flux, linking primary, secondary, windings
ω = angular frequency

BIBLIOGRAPHY

The following references apply to computer-aided design of magnetic circuits used in transformers and electric machines.

Transformers

S. B. Williams, P. A. Abetti, and E. F. Magnusson, "Application of digital computers to transformer design," *AIEE Trans. pt. III: Power Apparatus and Systems*, vol. 75, pp. 728–735, August 1956.

L. J. MacKinnon, "Power transformer design and estimate cost program with an IBM 650 digital computer," *AIEE Trans. pt. III: Power Apparatus and Systems*, vol. 77, pp. 1262–1266, February 1958.

S. B. Williams, P. A. Abetti, and H. J. Mason, "Complete design of power transformers with a large size digital computer," *AIEE Trans. pt. III: Power Apparatus and Systems*, vol. 77, pp. 1282–1291, February 1958.

W. A. Sharpley, and J. V. Oldfield, "The digital computer applied to the design of large power transformers," *Proc. IEE*, (London), vol. 105, pt. A, pp. 112–125, 1958.

L. Inkenen, "Die Optimale Transformerenberechnung auf Digitalrechenmaschinen," *Electrotech u Maschinenban*, Heft 19/20, pp. 464–469, October 1963.

T. H. Putman, "Economics and power transformer design," *IEEE Trans. Power App. Systems*, vol. PAS-82, pp. 1018–1023, December 1963.

P. P. Yeh and E. Cohen, "Leakage reactance of ring-type transformer with rectangular core section," *IEEE Trans. Power App. Systems*, vol. PAS-84, pp. 684–690, August 1965.

R. L. Perttula, "Transformers in one second," *Allis Chalmers Engineering Review*, vol. 32, no. 3, pp. 23–25, 1967.

Electric Machines

C. G. Veinott, "Induction machinery design being revolutionized by the digital computer," *AIEE Trans. pt. III: Power Apparatus and Systems*, vol. 75, pp. 1509–1517, February 1957.

G. L. Goodwin, "Optimum machine design by digital computer," *AIEE Trans. pt. III: Power Apparatus and Systems,* vol. 78, pp. 478–488, August 1959.

G. W. Herzog, O. W. Andersen, J. Scrimgeour, and W. S. Chow, "The application of digital computers to rotating machine design," *AIEE Trans. pt. III: Power Apparatus and Systems*, vol. 78, pp. 814–819, October 1959.

C. G. Veinott, "Synthesis of induction motor designs on a digital computer," *AIEE Trans. pt. III: Power Apparatus and Systems,* vol. 79, pp. 12–18, April 1960.

A. E. Hartman and G. V. Mueller, "Computer design of capacitor start motor windings," *AIEE Trans. pt. III: Power Apparatus and Systems,* vol. 80, pp. 1087–1090, February 1961.

L. M. Buchanan and T. F. Winters, "The design of single-phase induction motors using a digital computer," *IEEE Trans. Power App. Systems*, vol. 82, pp. 891–896, December 1963.

L. E. Welch, "Design of small D-C machines with a digital computer," *IEEE Trans. Power App. Systems,* vol. 82, pp. 1099–1106, December 1963.

K. J. Waldschmidt, "A computer procedure for single-phase induction motor calculation and design," *IEEE Trans. Power App. Systems*, vol. 82, pp. 867–875, December 1963.

D. W. Novotny and J. J. Grainger, "Digital computer analysis of the instantaneous reversal transient in a single-phase motor," *IEEE Trans. Power App. Systems*, vol. PAS-83, pp. 380–386, April 1964.

S. Lehman, "Analytical representation of magnetization characteristics for digital computation," (In German), *Elektronische Datenvarbeitung*, vol. 6, pp. 165–172, August 1964.

E. A. Erdelyi, S. V. Ahamed, and R. D. Burtness, "Flux distribution in saturated DC machines at no-load," *IEEE Trans. Power App. Systems*, vol. 84, pp. 375–381, May 1965.

M. S. Erlicki and J. Applebaum, "Optimized parameter analysis of an induction motor," *IEEE Trans. Power App. Systems,* vol. PAS-84, no. 11, pp. 1017–1024, November 1965.

O. W. Anderson, "Optimum design of electrical machines," *IEEE Trans. Power App. Systems,* vol. PAS-86, no. 6, pp. 707–711, June 1967.

H. E. Jordan, "Digital computer analysis of induction machines in dynamic systems," *IEEE Trans. Power App. Systems*, vol. PAS-86, no. 6, pp. 722-728, June 1967.

L. F. Wiederholt, A. F. Fath, and H. J. Wertz, "Motor transient analysis on a small digital computer," *IEEE Trans. Power App. Systems*, vol. PAS-86, no. 7, pp. 819–824, July 1967.

F. C. Trutt, E. A. Erdelyi, and R. E. Hopkins, "Representation of the magnetization characteristics of DC machines for computer use," *IEEE Trans. Power App. Systems,* vol. PAS-87, no. 3, pp. 665–699, March 1968.

INDEX